Quirky Quarks

The Authors

Benjamin Bahr, Boris Lemmer, Rina Piccolo

Dr. Benjamin Bahr is a quantum gravity researcher at the University of Hamburg, Germany. He and his research group work on a unification of Einstein's theory of general relativity with the principles of quantum physics.

Before that, he did his PhD at the Max-Planck Institute for Gravitational Physics in Potsdam, and was a research fellow at the University of Cambridge, UK.

When he is not trying to calculate what goes on inside a black hole, or what happened at the Big Bang, he likes to explain physics to laypeople – by giving public talks, or writing popular science books.

Dr. Boris Lemmer is an experimental elementary particle physicist, working at the University of Göttingen and on the ATLAS Experiment at CERN.

Before doing his PhD in Göttingen, he studied physics and mathematics in Gießen. He does not only love science, but also explaining it to laymen, either in books, in talks or on stage. In 2011, he won the German Science Slam championship.

Rina Piccolo's cartoons have appeared in numerous magazines including The New Yorker, Barron's Business Magazine, The Reader's Digest, Parade Magazine, and more.

Her daily comic strip "Tina's Groove" is syndicated in newspapers and websites worldwide.

Benjamin Bahr · Boris Lemmer ·
Rina Piccolo

Quirky Quarks

A Cartoon Guide to the Fascinating Realm of Physics

 Springer

Benjamin Bahr
Hamburg, Germany

Rina Piccolo
Toronto, Canada

Boris Lemmer
Göttingen, Germany

ISBN 978-3-662-49507-0   ISBN 978-3-662-49509-4 (eBook)
DOI 10.1007/978-3-662-49509-4
Springer Heidelberg Dordrecht London New York

Library of Congress Control Number: 2016932389

Managing editor: Margit Maly
Illustrator: Rina Piccolo
Cover Illustration: Rina Piccolo

Printed on acid-free paper

Springer Berlin Heidelberg is part of Springer Science+Business Media
(www.springer.com)

Preface
What Do You Get When You Collide Two Physicists and a Cartoonist?

If you remember your college, or high school physics textbook like I remember mine, you'll agree that it was the heaviest of all your books to carry, and the most difficult to understand. A quick glance through the pages of this book – the one you're looking at now – will show you that it looks nothing like your college, or high school textbook. It's way too fun looking. Flip through it and you'll find two dogs, a cat, and quarks with three eyes. You'll see cartoon electrons, and comics about Special Relativity. Oh, and let's not forget the actual science. Explained in plain, everyday language that even a cartoonist like me can understand, co-authors Boris Lemmer, and Benjamin Bahr will show you what anti-matter has to do with bananas, and why you feel bloated on an airplane (clue: it's not because you ate bananas). They'll crack open an atom, and make you question the stability of the ground beneath your feet. They'll tell you how to create matter out of energy. They'll even have you wondering about a possible other you, in a possible other universe, reading another book eerily like this one.

As a cartoonist and writer, I am a natural wonderer, and although like many people I find physics difficult, the stuff of black holes, worm holes, and sub-atomic strangeness has always intrigued me. The opportunity to work with Boris, and Benjamin – physicists devoted to de-coding nature's biggest puzzles – has made me a better wonderer, and a shade brighter. Thanks to these gentlemen, I now understand things like surface tension… and find it as mind-boggling as dark matter. (Thanks, guys, for adding to my crazy mental catalogue of things I love to wonder about.)

If you're like me, an avid wonderer who enjoys having your mind blown by the often bizarre nature of reality – and you like cartoons – then seek no other book than this one.

So, what do you get when you collide two physicists and a cartoonist? You get quarks, quirks, and an enjoyable exploration of the fascinating realm of physics.

–Rina Piccolo

Contents

II – The Cosmos 71

III – Quantum Mechanics 141

IV – Particle Physics 203

V – Beyond the Boundaries of Our Knowledge 269

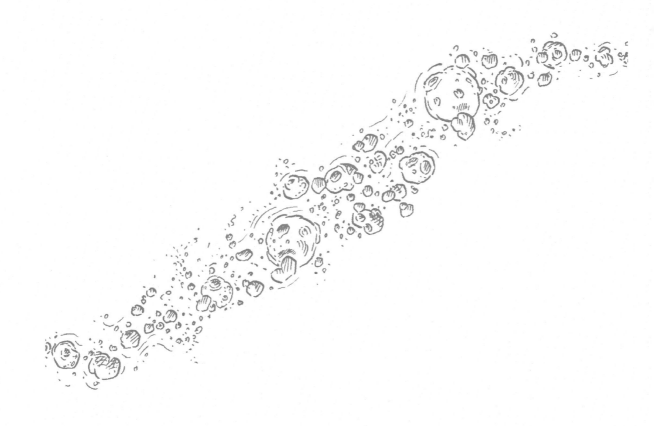

The CHARACTERS

ERWIN

He can explain the Uncertainty Principle, but is himself uncertain as to what to do with a hairbrush. He can find his way through a Feynman Path, but is hopelessly lost in a shopping mall parking lot.

That's Erwin – our theoretical physicist at Princeton. Only a brilliant guy like him can get away with applying the Many-Worlds Theory to laundry day. As he puts it, "I'm satisfied with the probability that, if not in this reality, then at least in some other alternate reality – my cardigan is being washed."

As a young pup in his dad's garage, he built rockets that touched the edge of space. No small wonder that Maxwell – his paws on switches and dials, his head in the stars – was destined to explore the universe.

Today, at MIT, where Maxwell spends his time, you'll most likely find him inside a lecture hall giving a spirited talk on cosmic voids, inside a lab tinkering with gadgets and screens, or in the local pub discussing wormhole navigation with colleagues over a pint. And oh, he has a special place in his heart for women mathematicians – perhaps one in particular.

MAXWELL

EMMY

The beauty of the night sky is as important to her as the accuracy of a mathematical proof. Behind Emmy's passion for numbers is, you might say, a personal quest to uncover the rational elegance in the natural world around her.

When she's not running computations in her office at Oxford, you'll find Emmy on a dinner date with Maxwell, or debating the existence of gravity mediating particles with Erwin.

The PARTICLES

Proton

If an atom were the size of a football field, this little guy – the Proton – would be smaller than a spider on that field. Made up of two UP Quarks, and one DOWN Quark, the Proton may not be an elementary particle, but his role in the quantum world is enormously important. It's this little fellow that makes up every nucleus in every atom in the vast universe. He's a happy particle, being always positively charged!

Electron

Most of the time, you'll find this negatively charged particle spinning in his home shell – or orbit – inside an atom. We've got a lot to thank him for. Think about it – if it weren't for him and his pals we would not have electricity. With his truly magnetic personality, you'll rarely find the Electron alone, but always seeking other particles, and fellow Electrons, to hook up with. It's this binding and arranging with his clan that determines all chemical reactions in the universe!

Never was there a more charming team of particles than this one – the team of Quarks. There are – as far as we know – six players in the Quark lineup, named Up, Down, Charm, Strange, Top, and Bottom. With whimsical names like these, you'd think Quarks are the silly clowns of the subatomic world – and you'd be wrong. As elementary particles that cannot be further broken down, these characters are the building blocks of several other particles, like Protons, Neutrons, and Hadrons. Silly clowns they are not!

Neutrino

If you are not well acquainted with the Neutrino and his kin, then you should be. He and his trillions of friends pass through your body every second of every day. But don't worry – although these intrusive little runts are not yet fully understood by physicists – they can't hurt you. It's a good thing that the Neutrino has a mass of almost zero, and rarely interacts with objects made of normal matter like your dog, your cat, your aunt, or your aunt's hairdo. The Neutrino comes in three "flavors": Electron Neutrino, Muon Neutrino, and Tau Neutrino. But don't put one in your mouth because – oh, wait… you most likely are chewing on a bunch right now.

Don't let their nickname "Weak Bosons" fool you. These massive folks are the heavyweights in the land of elementary particles. Still, it's hard not to feel a little sympathy for them – with a lifespan millions of times shorter than the lifespan of a fruit fly, W and Z are only around long enough to do their job, and then they bite the dust. What's their job? Not a small one. These Bosons are responsible for mediating one of the fundamental forces of nature, the Weak (Nuclear) Force, acting as force-carriers between other particles. It's W and Z that we should thank for keeping us alive, for without the Weak Force, the sun would not be able to burn, and shine its light on us!

W/Z Bosons

Higgs Boson

Possibly the particle world's biggest celebrity, the Higgs Boson was only just recently discovered, and has been the focus of headlines around the world. But the Higgs didn't always live a life of fame – for decades, he enjoyed a game of Hide-And-Seek with physicists, and made a teasing presence on the Standard Model of Particle Physics. Why all the attention? Let's put it this way: this little guy makes matter massive! Imagine for a moment a universe made of massless objects. No relaxing walks in the park, no football games, nothing. We would all just move at the speed of light! How does Higgs do it? Well, we know that he hangs out in the Higgs Field, and messes around with the Weak Force, and the Electromagnetic Force, giving other particles their mass. It would not be unreasonable to say that the Higgs Boson is more than a celebrity; he's a major player in the symphony of the universe.

Measure for Measure
On the Units in Science

Physics is the natural science of the behavior of (non-living) matter, and how they interact with each other. Well, some of that is also covered by Chemistry, but the boundaries between these two areas of science are rather fluid anyway. To understand the matter of our universe, one needs to measure it. How much is there? How heavy is something? Where is it? How large? How long did it take? These questions can be answered by measuring – in some cases requiring a sufficient budget, of course (↗1).

The results of our measurements always come in numbers, and they have a unit. So for instance, the process of drinking a good beer could take you 7 minutes. In that case the "7" would be the corresponding number, and "minutes" would be the unit. You could also have used another time unit, such as seconds. Since 1 minute is 60 seconds, you could have also said that the same beer took you 420 seconds to finish. These two statements are completely interchangeable. Whatever you use is just up to you, and your – and other people's – convenience. Although it might be not very useful to claim that for finishing your beer you needed only 0.1339 millionths of a century, you would be correct in that statement.

Now throughout human history, a plethora of different measurement units have been used. In the scientific world, a few have been decided to form the basis of measurements, in which we express all physical quantities. These form the so-called Système International d'Unités (SI). The three most important SI units are units for length, time and

mass: the meter, the second and the gram (As a side remark: there are four more, the Candela (Cd) for brightness, the mole (mol) for an amount of a substance, the Kelvin (K) for temperature and the Ampère (A) for electric current, of which only the Kelvin will play a role in this book.). All other used units for length, time and mass are derived from those by the proper prefix: kilo (k) for thousand, mega (M) for a million, giga (G) for billion, and so forth. Fractions of units are denoted with a similar prefix: milli (m) for a thousandth, micro (μ) for a millionth, nano (n) for a billionth, and so forth.

This means that we can express all physical quantities in their specific units, and it is very easy to change between units – simply shift the decimal: 625 nanometers are the same as 0.625 micrometers, and so on. But, depending on where you are from, these might not be the units you are used to. How much is that again in inches, or ounces? When writing a book about physics, this posed a problem for us: do we use the scientifically correct units? Or do we use the ones which everybody understands easily?

In the end, we have decided to go the middle way. For most physical quantities we use the SI units … unless we don't. Because even European scientists and strong proponents of the metric system (both of which the authors confess to be) occasionally use non-SI units in their physical articles. Do we use mega-and gigaseconds? Of course not, we use years and centuries in our research. Instead of peta- or exameters we use light years and parsecs.

↗1: "Particle Accelerators" on page 249

And in nuclear physics everybody uses Ångström, which equals 0.1 nanometers (because let's face it: Swedish umlauts rock!).

Moreover, for quantities which are in the realm of our everyday life, it seemed silly to enforce the metric system, when ordinary miles and inches work better. Indeed, for the purpose of this book they are the better choice, because they give an immediate idea of the dimensions involved, without forcing the reader to always calculate from the metric system to the US American one. For very large or small numbers, however, we revert back to SI units (with the exceptions hinted at earlier), because not only

are there no nanoinches, but also to stay in line with most of the physics literature.

We have tried to jump back and forth between the two systems as little as possible, but wherever necessary. We hope that this will not cause confusion, but add clarity. If all goes well, you will not notice it at all during reading. In fact, there are only very few articles where we explicitly state a quantity in both US and SI units.

So please enjoy the science – in whichever unit you prefer! And if you are confused, just visit one of the many unit converters, as in ➚[2].

➚[2]: http://www.amamanualofstyle.com/page/si-conversion-calculator

I – Rocket Science

"Science is amazing! With the power of science we can fly, cure diseases, predict earthquakes and solar eclipses, and put the fizz in soda. Occasionally, something blows up, but hey – where is the harm in some spontaneous pyrotechnical experiment? It's just science in action!

And the science you can make with just physics – wonderful! I could work and tinker all day, until I have built a new, awesome piece of technology. I love it! There is just so much you can do, just with physics alone: lasers, flying machines, invisibility cloaks (seriously!), and rockets. Oh, the rockets! They are my favorite! Every once in a while, I give this pastime a go. It is an old dream of mine: traveling through space in a rocket, visiting other planets, taking close-up pictures from asteroids, reaching the outer edges of the solar system... sigh.

On the next few pages, I'll show you some of physics' finest! How airplanes fly, how lasers work, and which way the Voyager probes took through the solar system. I hope you'll find it as exhilarating as I do!"

Auroras
An Exciting Glow for Humans and Atoms

Video games and movies have developed rapidly and are able to impress us with their visual effects. But fortunately, nature is still the undisputed number one. To see that, just travel to one of the very northern or southern parts of Earth, wait for a dark night and look up in the sky. If you are lucky, you can see one of nature's most impressive phenomena: an aurora. They have different appearances, but look all more or less as if the dark sky turned into a giant green glowing soup that somebody is stirring with a giant, invisible spoon.

It really looks like a miraculous effect. The Vikings thought it was a sign of a great battle that had just occurred somewhere on Earth, and also other cultures interpreted auroras as a sign from supernatural beings. But what is actually going on up there? Why are they green, show up only from time to time and can be seen only far up in the north ("Aurora Borealis") or in the south ("Aurora Australis")?

Wind from the Sun

A first hint comes from the fact that you need good weather to see them. Not only does the sky have to be clear, it is also another kind of weather that has to be just right. It is called "space weather" and is concerned with changing conditions up in space, such as moving particles and magnetic fields ($\nearrow^1$). If you take a look at space weather status reports and forecasts ($\nearrow^2$) you will see that one of the main indicators for space weather is the solar activity. If you watch the sun (please, don't stare directly into

it!) you can see dark spots on it. These are places with very strong magnetic fields. From time to time the sun will eject a big bunch of particles from these dark spots: protons, helium nuclei (as a fusion product, $\nearrow^3$) and electrons. Such phenomena are called "coronal mass ejection". Their number depends on the activity of the sun, which increases and decreases in cycles ($\nearrow^4$). You can say: the more spots, the more mass ejections. The ejected particles will then travel through space, some of them maybe pointing towards the Earth. We call this the "solar wind", which is composed of both the bursts as well as of a constant flow of particles that leaves the sun continuously.

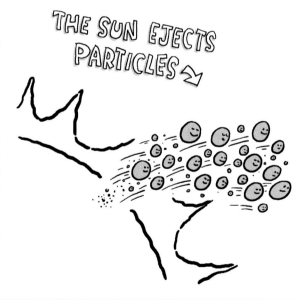

$\nearrow^1$: *"Light" on page 7*
$\nearrow^2$: http://www.swpc.noaa.gov

$\nearrow^3$: *"Nuclear Fusion" on page 175*
$\nearrow^4$: https://en.wikipedia.org/wiki/Solar_cycle

This solar wind is part of the space weather. Whenever you talk about weather you automatically want to get a forecast as well. While on Earth you could complain about the forecast quality quite easily, the space weather forecast is pret-

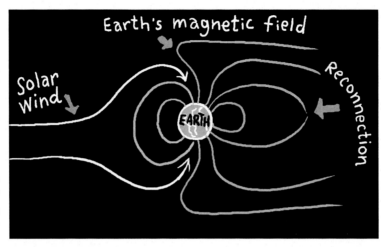

ty reliable. The reason is that you will exactly know that something will arrive from the sun because you already see it happen while the weather is still on its way. This can be as the visual information from the sun reaches us with the speed of light (about 670,000,000 mph), while the solar wind with its massive particles is much slower (about 900,000 mph).

Magnetic Field Crashes

Imagine a strong solar wind reaches the Earth. We can be quite happy that good old Earth has both an atmosphere and a magnetic field. Such a particle wind is nothing but radiation that would seriously harm us if we would be directly exposed to it. Fortunately, our atmosphere can absorb this radiation. But even better than that, the Earth's magnetic field deflects the solar wind! It will be deflected along the magnetic field lines which are shaped as indicated in the illustration. The solar wind itself changes the shape of the Earth's magnetic field. As a bunch of moving particles, the solar wind is nothing else

but an electric current. And currents cause magnetic fields. So the magnetic field lines push from the sun's side, squeeze the magnetic field on the left and drag it to the right. There are several ways that the particles from the solar winds can enter Earth's atmosphere. One is via the open field lines at the "polar cusps", where the Earth's magnetic field cannot protect us. Another way is via the tail in the right. Squeezing the magnetic field from the top and bottom on the right side can make the field lines reconnecting and flipping back towards Earth. The more solar wind, the more the Earth's magnetic field lines are bent. Heavy changes are also called "geomagnetic storm" (↗5). After reconnection, the field lines take a bunch of solar wind with them. These particles can also push others out of the so-called "Van Allen radiation belt" (↗6) and push them into the atmosphere. This belt is a place of charged particles trapped in the Earth's magnetic field, constantly fed via the solar wind and other intergalactic sources of particles.

Particle Rain and Glowing Atoms

Now we have charged particles entering the Earth's atmosphere at the place of magnetic field line gap close to the North and South Pole. While the particles are travelling, they hit the atoms of the Earth's

↗5: https://en.wikipedia.org/wiki/Geomagnetic_storm
↗6: https://en.wikipedia.org/wiki/Van_Allen_radiation_belt

An OXYGEN ATOM is hit by an electron:

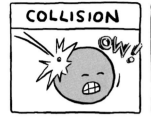

COLLISION

OW!

GETS EXCITED

GA!

STABILIZES QUICKLY VIA EMISSION OF GREEN LIGHT, OR SLOWLY VIA EMISSION OF RED LIGHT:

OR

atmosphere. During such hits they transfer energy to the atoms and bring the electrons to an "excited state". From this state they can relax again via the emission of light. Depending on the energy difference of these states, the wavelength of the emitted light varies from red (low energies) to blue (high energy) (↗[7]). Oxygen atoms emit light in the visible range, dominantly in two ways. One corresponds to red, the other one to green. The time it takes the electron to move from an excited to a ground state is different: the green transition happens within less than a second, the red one takes longer. If we are far up in the sky, the atom density is very low. Oxygen atoms are alone and take their time to emit red light. This red emission happens much more often than the green one. Closer to the surface, the atomic density increases, the atoms collide with others, transfer energy via collisions instead of light emission and the red light is suppressed. What remains is the "fast" green light. At even lower altitudes, the density of atomic oxygen is so low (not to confuse with oxygen molecules whose density increases with lower altitude) that we no longer see any light emission. This is the reason why auroras are red far up (not always visible) and then mostly green.

Now, take your time to look at an aurora. Isn't it beautiful? Certainly an experience you wanted to share, such as Erwin and Maxwell.

↗[7]: "Light" on page 7
Image: Hugo Løhre / NASA

Light
Ripples in the Electromagnetic Field

What is light, precisely? Well, light is a specific kind of electromagnetic radiation. By that we mean "ripples in the electromagnetic field".

The Electric Field: To Charge or Not to Charge, That is the Question

Let us first explain the electric field: Well, you know that things can be electrically charged, either positively or negatively. Things with similar charges repel each other, while things with opposite charges attract each other. So, electrically charged matter always feels a force, coming from all the other charged things in the universe. This force field (physicist call this a "field" because it is everywhere) is called the electric field. You can think of it as the GPS for charged particles – it tells them where to go next.

Magnets – How Do They Work?

The magnetic field is a bit more difficult to understand. Things can also carry a magnetic charge, we then call them "magnetized". But the magnetic charge is not just a number, like the electric charge. Rather, it has a strength, and at the same time also a direction. You can think of this as a little arrow attached to the magnetized object – the arrow points away from the North Pole and towards the south pole of the magnet, and the longer the arrow the stronger the magnet. Just as electric charges try to push charged particles around, so do magnets. But while electric charges exert a force on charged particles simply because they are charged, magnets only act with a force on them (the so-called Lorentz force) if they move. So as long as a proton sits perfectly still, a magnet cannot push it around. Only when the proton starts to move somewhere – maybe because of the electric field – will the magnets be able to change its path.

Two Sides of the Same Coin: The Electromagnetic Force

The magnetic force has been known for thousands of years, as long as people realized that needles made of certain metals always point into the same direction – the North Pole. The ancient Greeks also knew about the electric force. They observed that by rubbing amber on animal fur caused little crackling sparks (which we nowadays call "electric discharges"). In fact ηλεκτρον, or "ēlektron", is the ancient Greek word for amber.

But it wasn't until the end of the 19th century, when English scientist James Clerk Maxwell found out that the two belong together. They are just two different phenomena of the same physical interaction, which we nowadays call "the electromagnetic field". The electric and magnetic force belong together inseparably. The famous equations which made J.C. Maxwell immortal in fact state that a changing electric field causes a magnetic field – which is why a spinning electric charge behaves like a magnet (↗[1]).

↗[1]: "Spin" on page 187

But conversely, a changing magnetic field causes an electric field – which is why a rotating magnet is used in every kind of power plant to push electrons through wires (in other words: generate electricity).

So if you disturb the electric field, this generates a change in the magnetic field, which in turn again changes in the electric field – in the opposite direction of the initial disturbance. This is also called "Lenz' rule". For the magnetic field precisely the same is happening, which is why the electromagnetic field is a bit like an elastic rubber sheet: if you poke at it, it wants to get back into its original form, if possible. But this going back to its original form doesn't happen right away, but takes some time. Which is why a disturbance in the electromagnetic field – like a ripple in the rubber sheet – will spread out, oscillating back and forth. These wave-like ripples propagate in all directions with a very specific speed. They are called "electromagnetic waves", and the speed is the well-known speed of light.

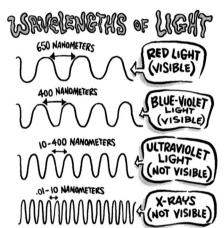

The Electromagnetic Spectrum: Size Does Matter

Although the speed of these waves is always the same, the frequency with which they make the electromagnetic field oscillate, can be different. The faster they jitter back and forth (and thus, the shorter their wavelength), the more energy is stored in them.

The longest waves are the long radio waves. Their wavelength is above a kilometer (i.e. roughly a mile). The shortest of those are the waves we actually use in our radios: AM waves are a kilometer up to a hundred meters long, while FM waves are about one to ten meters in length. The next shorter waves are called microwaves, and their wavelength is from a meter down to a millimeter. The ones in your microwave oven are actually about 12 cm, or 4.7 in. This is not yet very energetic at all, but coincidentally it is precisely the energy of a quantum transition in water molecules, which is why these waves are ideal to heat up anything that contains water.

The electromagnetic waves from about one to about a hundred micrometers are called infrared (IR) radiation. This radiation is not visible to the human eye, but we can feel it, as being warm. Put your hand next to an oven or a roaring fire, and you can feel the infrared radiation "first-hand".

The IR ends where the visible light begins – the electromagnetic radiation from 700 nanometers (red) to about 390 nanometers (dark blue) is the visible light. Our human eyes are made so we can see this radiation. It is what we call light, and consists of those colors which you can see in a rainbow – ordered from red to blue.

We now come to more and more energetic waves. Beyond the color blue in the electromagnetic spectrum is the ultraviolet (UV) radiation. Their lengths range down to about 10 nanometers, which is where the X-rays begin. X-rays have a wavelength down to 10 picometers (a trillionth of a meter). Although it can be quite useful in medical diagnostics, this kind of radiation is already so energetic that it can be dangerous if one overuses it.

And finally, everything with an even shorter wavelength than picometers is called γ-rays. This extremely high energy radiation reaches us from outer space, when stars explode (↗2), or when heavy atomic nuclei decay (↗3).

So It's a Wave – or Is It?!

From earliest times natural philosophers wondered about the nature of light. For quite some time, an idea that people had was that light consisted of tiny little particles, which were flying from objects into our eyes. However, its wave-like properties were understood around the 18th century, when natural philosophers discovered that light, just as water waves, can be refracted, aberrated and scattered. And with J.C.Maxwell in the 19th century, it became finally clear that light was a certain type of electromagnetic wave. The question of the nature of light seemed to be settled, and the idea of little "light particles" was finally buried. At least, this was what people thought these days.

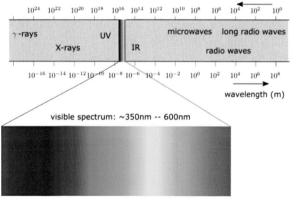

visible spectrum: ~350nm -- 600nm

It was around 50 years after J.C.Maxwell's ideas, that a young scientist made a remarkable discovery: He found that one cannot put arbitrarily little energy into an electromagnetic wave. Usually, the energy which is carried by it is determined by two factors: one is its wavelength, and the other one is its intensity, or its brightness.

Now this young scientist – his name was Albert Einstein – found that you cannot have an arbitrarily low intensity within an electromagnetic wave. Even more, the energy in a wave can only be transferred to other systems in packages of a fixed size, or quanta. These energy quanta are larger, the shorter the wavelength is. In other words, there can be two electromagnetic waves with the same energy – one very bright one with a long wavelength, and one very dim one with a short wavelength – but in the first one there are many energy packages of small size, and in the second one there are only few packages, but each containing a lot of energy.

These energy packages are called photons. They can be thought of as tiny little particles which light (or any other kind of electromagnetic radiation) consists of. Einstein later received the Nobel prize for this discovery, and this was the first clear example that a physical system can behave both like a wave and a particle. This is why Einstein is sometimes mentioned as one of the inventors of quantum theory (↗4).

↗2: *"Supernovae" on page 83*
↗3: *"Alpha, Beta and Gamma Rays" on page 171*
↗4: *"Wave-Particle Duality" on page 143*

Invisibility Cloaks
Walk Like a Magician

It is an old dream: walking through the streets without being seen. Just put on a magic cape, and become invisible. One would be able to pull the best practical pranks ever! Although the most purposes this invisibility cape would be used for would quite probably not be very nice, probably everybody has thought about it at some time in their life.

Now You See Me – Now You Don't

This is why the world listened up, when researchers of Duke University in North Carolina announced 2006, that they had built a functioning cloaking device! The first enthusiasm was slightly curbed, however, when they added that it would only work for small objects, only for microwaves, and only when looking from a very specific direction. So, instead of turning an object invisible to the naked eye, it would allow for it to be placed in a running microwave oven without getting hot. Still, it is a good start. But how does the "invisibility cloak" work, precisely?

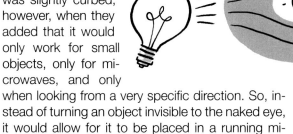

Bending Light – the First Step to Invisibility

Before we can understand how invisibility works, we need to be clear on what it is that makes us see things. The ancient Greeks assumed that our eyes worked like ray emitters, sending out some kind of sensory beam. Whenever that beam hit something, we would see it. Nowadays we know that it is pretty much the other way round: A light ray from the sun (or another source) hits an object. Of all of those rays hitting it, some are absorbed (heating it up), and some are reflected, depending on the wavelength of the light ray (↗[1]). The reflected light rays, whenever they hit our eye, give us the impression of a certain color (depending on the wave length) in that direction.

So the theory is easy: For an object to become invisible, we just need to prevent it from getting hit by light rays! So an invisibility cloak should simply guide the light around the object it envelops. Whenever a light ray hits the cloak, it should not reflect the beam into any direction, but should alter its path so that it can emerge on the other side, as if there had never been an obstacle at all.

Refraction – Can It Help?

All right, so for an invisibility cloak to function, one needs to be able to bend light. Materials can do this, it is called refraction! Refraction happens every time light passes from one medium to another – it slightly changes its direction. This is why it is so dif-

↗[1]: "Kepler's Laws" on page 47

ficult to catch a fish with your bare hands: the fish is not at the position where you see it, but slightly next to it. The reason is that the light gets refracted, when it passes from the fish, through the border between water and air, to your eyes. So can one use this to make the fish completely invisible?

Now here comes the big problem: The angle under which something is refracted depends on the so-called refractive index of the material it passes through. Vacuum has a refractive index of precisely 1, air has an index from ever so slightly larger than 1, water has one of 1.33, glass of about 1.5, and so on. Normal materials in the world all have a refractive index equal to or larger than 1! Because of this, light can only be bent towards the object, never away from it. For guiding a light ray around an object with refraction, one would need a material which at

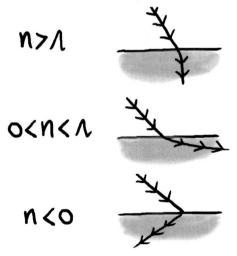

some points has a refractive index of smaller than 1, or even negative indices, which are less than zero!

The Refractive Index: the Crux of the Problem

Materials with these refractive indices would have very peculiar properties. If the index would be between 0 and 1, light would not only be bent away from the object, it would also travel faster than the speed of light. And in materials with negative refractive index, light would even travel backwards! Would that even make sense?

Metamaterials: Bending Light to Our Will

But this is precisely what the researchers at Duke University have achieved: to create a material which has, effectively, a refractive index smaller than one. How did they manage that? Well, the optical properties of the material do not come from the types of atoms it consisted of, but also from the way these atoms were arranged: They built a ring, made of arrays of lots of tiny copper structures, only micrometers wide. Whenever a light wave with a certain wavelength would hit these little structures, they would start to emit electromagnetic waves themselves, like thousands of little antennas. Because of constructive and destructive interference, these little waves would add up, and the resulting light wave would be traveling in a different direction than the initial one. The changed direction would be such that, effectively, the material behaved as if it had a refractive index of less than one.

Because the interference of electromagnetic waves is very dependent on the actual wavelength, it should be no surprise that it only works for waves with a very specific frequency – in this case, microwaves. But it did work: If the ring was hit by microwaves from a certain direction, the waves would be guided around the ring, and emerge on the other side, as if nothing had obstructed their path. Any object placed in the center of the ring was completely shielded from the microwave radiation.

These materials, whose optical properties are specifically designed by their microscopic structure, are also called meta-materials, and research in these has exploded in the past decade. Soon, versions of the "invisibility cloak" were produced, which worked with larger objects, moving objects, sometimes even for some parts of the visible spectrum.

One could easily think of other uses: By constructing the right kind of barriers, one might be able to guide water waves around small islands. Possibly quite useful for those little archipelagos which are frequently in danger of drowning in storms. Although sturdy houses and good raincoats might actually be cheaper…

However, after all is said and done, we have to conclude that, at the time this book is written, no cloak has been constructed which would make us actually feel like Harry Potter. Although there is a lot of research happening in this area at the moment, we are still some major steps away from being able to walk through the streets without anybody being able to see us. But who knows what the future will bring?

"Invisibility" Also for Other Types of Waves

Because what is used here is essentially the wave-like nature of light, one could ask oneself: does this also work for other types of waves? Of course it does! Just a few years ago, the researchers at Duke University presented a "silence cloak". This was constructed out of 3D-printed materials, and was able to shield an object completely from sound waves. You can see it in the picture on this page. A person covered with such a silence cloak would hear nothing from the outside: sound waves would be guided around it completely. Bats would fly right against them – they'd be completely "invisible" to their sonar.

Image: Reprinted by permission from Macmillan Publishers Ltd: "Three-dimensional broadband omnidirectional acoustic ground cloak", Lucian Zigoneanu, Bogdan-Ioan Popa & Steven A. Cummer, Nature Materials 13, 352–355 (2014)

"I wasn't speeding. I was demonstrating to my colleagues how police radar employs the principles of the Doppler effect."

The Doppler Shift
The Stretching of Waves

What is the sound of a car engine? Well, actually, that depends very much on where you are! If you sit inside the car, and have time to listen to the engine, you'll hear a constant humming of the machine. Quite a muffled sound, actually, because nowadays car manufacturers try to make sure that you hear as little of the engine sound as possible.

But if you sit by the side of a road, then a passing car sounds quite different. If it is approaching, the sound will be quite high-pitched, but at the moment the car passes you by and recedes in the distance, the sound will switch from high to low pitch. Wreeeeeeeerooouuummmm! If you have ever hitchhiked, you will have heard that kind of sound a lot! The reason why you hear two different pitches of sound is the so-called Doppler effect. It happens with all kinds of waves!

But with sound waves it can be explained most easily, because sound waves are just rhythmical compressions of air. If there is total silence, then the air pressure (↗¹) is the same everywhere. But as soon as, at one place, the pressure is a bit different from other places, this pressure difference will wobble through the air, spreading out in a wave-like manner. It is this pressure wave which we perceive as sound, as soon as it reaches our ear. And the shorter the wavelength, the higher the

frequency, and the higher the pitch of the perceived sound.

Vibrating Machines and Their Sound

Why does a car engine make a noise, after all? Well, such a machine has a lot of movable parts, and when an engine is running, many of these parts are vibrating. If such a piece of, say, vibrating metal, has contact to the air, it will make the air vibrate, too. Well, it will compress the air around it a little bit every few fractions of a second.

What is reaching our ear when we hear a sound is a succession of high pressure – low pressure – high pressure – low pressure – and so forth. As we have mentioned, we hear a higher pitch the shorter the wavelength, in other words, the faster the air changes between high and low pressure. Normally, if a machine vibrates with a certain frequency, then we'll hear a sound with exactly the same frequency.

Now, what happens when the machine is moving?

Moving Sound Waves Get Stretched – and Thus Pitched

What happens is that the sound wave does not leave it in every direction with the same frequency. Rather, in the direction in which the engine moves,

↗¹: *"Vacuum and Air Pressure"* on page 23

the waves are compressed. In the opposite direction, they are stretched. So if a noisy machine approaches us, it pushes the sound waves in front of it together. What then reaches our ear is a wave with a smaller wavelength – and that means a higher frequency, which in turn means a higher pitch. On the other hand, if the machine moves away from us, then the air waves that hit our ear are stretched slightly – so the wavelength is larger, and thus the frequency lower. We then hear a lower pitch.

know that this Doppler effect also happens for all other kind of waves – in particular for light waves ($\nearrow$[2])? And even better, using the Doppler effect one can actually measure the temperature of the sun, or how fast it rotates!

Well, one can if one has very precise measurements. You see, the sun in general is quite hot, and a lot of electromagnetic radiation reaches us from there, having all kinds of different frequencies. That

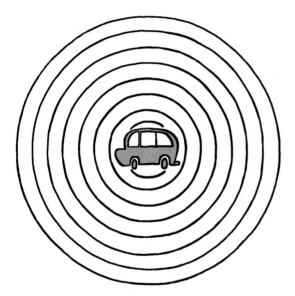 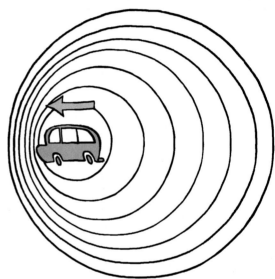

And if the machine passes us by? Well, then we hear the typical sound of a car engine, or a siren rushing past us: the pitch of the sound goes from higher to lower the moment it is on level with us.

The Doppler Effect for Light

The shifting in sound is probably something that has been experienced by almost everybody. But did you

is why the sun seems nearly white to us: its light contains all the colors of the spectrum.

If one looks very closely, however, one can see that there are some very specific frequencies which are missing! These are the so-called Fraunhofer lines, and physicists have known of their presence more than a hundred years before they understood the reason for their presence (or rather absence). Some

$\nearrow$[2]: *"Light" on page 7*

of the frequencies are missing, because they correspond to photons which have precisely the right energy to initiate a quantum leap of energy in the hydrogen (and other) atoms in the sun. For light with these frequencies, the plasma in the sun is completely opaque, while for all other colors of light it is basically transparent.

How Hot Is the Sun?

But now imagine that the hydrogen atoms in the sun don't all just sit there – they move! And they move quite a bit, the hotter they are. While a resting hydrogen atom needs to be hit by a photon with precisely the right frequency in order to absorb it, it can have a slightly lower frequency if the atom is, just in this second, moving towards the radiation. Because of the Doppler effect, a moving atom will see a photon with a slightly higher "pitch", which means a little bit higher energy.

What does that mean? It means that in the spectrum of the sun, there are not just precisely those frequencies missing which are responsible for the quantum leaps in hydrogen atoms – but also the frequencies slightly lower and slightly higher. They are also blocked out because of the Doppler effect! One says that the Fraunhofer lines have a certain width, and this width gets bigger the hotter the sun is – the higher the temperature the faster the hydrogen atoms, the bigger the Doppler effect.

And the Fraunhofer lines cannot just be made wider, they can also be shifted! You see, the sun is not fixed, it rotates around its axis. At its equator about one revolution in 25 days. That means that if you look at the light coming from the right edge of the sun (↗³), the Fraunhofer lines will be shifted to lower frequencies, while in the light from the left edge has Fraunhofer lines shifted to higher frequencies. And that happens, again, because of the Doppler effect: light coming from those parts of the sun which move away from us has lower frequencies – it's like the car moving away from us. One calls this light red-shifted. Consequently, light from the other edge of the sun is called blue-shifted.

Sgr A*: A Supermassive Black Hole – Proven by the Doppler Effect

The Doppler effect was also used in order to prove for the first time that there has to be a black hole (↗⁴) in the center of our Milky Way. You see, in the central region of our galaxy there are lots of stars orbiting a common center. We know this because we can see the microwave radiation from them arriving at the Earth. And because some of the radiation is blue-shifted and some is red-shifted, we also know how fast they orbit the center. And, people have found in the seventies of the last century, they rotate so fast that there needs to be an enormous mass which they all move around – such a great mass, in fact, that no normal star would be stable under that weight.

Nowadays we know much more about the supermassive black hole in our center – also called Sagittarius A* – in particular that it is more than four million times as heavy as our sun. Now *that* makes for some Doppler effect!

↗³: *"Nuclear Fusion" on page 175*
↗⁴: *"Black Holes" on page 91*

Lasers
High Quality Light Offering new Possibilities

There are flashlights, and there are lasers. They are somehow the same: devices that emit light. But we also know that there must be a substantial difference between the two. There is something that makes the mood become futuristic as soon as a laser turns up in a conversation. It is impossible to think of a modern, technological world without lasers. They read our CDs, DVDs and Blue Rays, can fix vision defects in our eyes, measure distances, read bar-codes and replace wooden pointers during presentations.

Atoms, the Light Factories

There are different ways of "producing light". A well-known mechanism is the temperature radiation. If you heat a material, it will emit light in all different colors. Not all colors are emitted in equal amounts. The color – or wavelength – distribution depends on the temperature. For low temperatures you might see no light at all, because the major portion of the emitted light Is in the infrared (↗¹). If it gets warmer you can see the object glowing red, than maybe yellow and finally even white (in fact, you see many colors in that light, but some dominate the intensity). Classical light bulbs emit such light via a heated tungsten wire.

But next to the classical light bulbs there are also devices that emit only light with a specific color. Take a sodium-vapor lamp, for example. It glows only yellow. The light works the following: a high voltage is applied so that an electric current begins to flow through the sodium gas. The sodium atoms get kicked by the electrons from the current and get energy transferred to them. As described by the laws of quantum mechanics, the sodium atom can absorb a certain amount of energy from the current's electrons and use it to bring one of its own electrons to a so-called "excited state". This means it has more potential energy than it had before. It is as if you would pick a ball from the floor and put it onto a table. You needed some energy to perform that move and the ball gained some. Your ball might roll down the table and release the energy again. And so will the sodium electron do. It will release the energy difference between its ground state and the excited state via the emission of a photon, a light particle. And these energy differences are specific for each atom type; the energy of the released photon will be as well. And as energies of photons correspond to the wavelength of

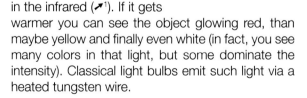

↗¹: *"Light" on page 7*

light, the wavelength will be different as well. And for sodium, the back-dropping electrons will emit yellow light.

In Lockstep, Guys!

All the photons emitted from the back-dropping elec-trons form the emitted yel-low light. There is one thing to notice: Each wave has a wavelength and a frequency that characterizes it. But each wave also has a certain phase. A phase defines the place at which the amplitudes are min-

imal and maximal. If two waves have exactly the same phase, they can interfere constructively and add their amplitudes. If you have two waves with the same wavelengths, but with a phase differing by half the wavelength, the two waves will cancel (read more about such interference in the chapter about the double slit experiment, ↗2). If all the photons of your sodium light have a different phase, you wouldn't care. But you would care if you wanted to focus the light. Too many different phases cause trouble and limit the amount to which you can focus your light.

The physicist's source of light is one where all photons have exactly the same phase, as if they would be in lock-step. And that's exactly what makes the differ-

ence between a classical light source and a laser! A laser can produce light with all photons in phase. We call this "coherent" light. But how does it man-age to do so? There is a way to allure an electron from an excited state to step down and emit light at a very certain time. This can be achieved if you don't simply wait until it does it on its own, but if you "stimulate" it. If an electron is in an excited state and a photon passes which has exactly the same wavelength as the electron would emit, the electron will step down! And it will do it exactly when the other photon passes by. So we have two photons which are – and here we have the specialty – exactly in phase. This principle we have to follow: Using photons to emit other photons which will then be in phase.

Building a Laser

What do we need? We need a material for which we can excite electrons. Somehow we have to manage to bring all electrons in these excited states and let them stay there for a while. Remember: they should not start falling back down automatically, but we want to induce this step with another photon. Typically, laser materials have more states than just a ground state and an excited state, but for our ex-planation these two states are sufficient. The first thing we have to do with our laser material is to "pump" it. Energy has to be injected that brings all electrons into the excited state. As soon as a first electron falls back down, the emitted photon will pass by another atom. And there it will make the ex-

↗2: "The Double Slit Experiment" on page 147

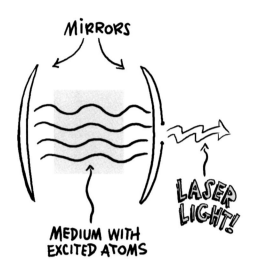

MIRRORS

LASER LIGHT!

MEDIUM WITH EXCITED ATOMS

wavelength, allow to resolve smaller structures. This is why we need blue laser rays to read out Blue Ray discs with smaller structures and hence more data. Focused laser light is also used to measure distances. Not only within your room or between buildings, but even to the moon. That's right, astronauts placed mirrors on the surface of the moon to allow us to measure the reflected light and to calculate the distance to the moon with an accuracy of millimeters.

cited electron fall down as well, releasing a second photon, in phase with the first one. And as these two keep on traveling, they will cause more photons to be released – all in phase! The material has to be surrounded by two mirrors that reflect the photons back and forth so that they can keep driving the avalanche of photons. Okay, at some point you also want to release all of your nice coherent photons out of that mirror box. Placing a small hole on one of the mirrors (or making the mirror partially transparent) will allow the laser light to shine through.

There are many different ways to realize such a laser. You can use a gas as active material (like CO_2), semiconductors, a ruby crystal, and many more. The material determines the wavelength of the laser light and hence its color. Many lasers are even out of the visible spectrum. As we mentioned in the beginning, there is a broad variety of applications for laser light. Its ideal focusing allows for example to read out DVDs with red lasers, Blue Ray discs with blue lasers. The blue lasers which have a shorter

Lasers exist not only in different colors but also with different laser pulse lengths (down to a few femtoseconds) and different powers. High power lasers can even be used to induced nuclear fusion (↗[3]) or as weapons. Another application is the production of artificial star light, projected in the atmosphere and used to correct the light from real stars for its disturbance by our atmosphere. The PARLA laser, shown in the photo, is used for such purposes by the European Southern Observatory in Chile.

↗[3]: *"Nuclear Fusion" on page 175*

Vacuum and Air Pressure
Molecules on the Move

Some things are so ubiquitous, we forget that they are there at all. Air is one of those things. We breathe it, and it keeps our bodies from popping (and freezing). But the instances we are aware of the air around us, are usually those which involve the air pressure – or more specifically the change of it. Let it be the whirring of the vacuum cleaner, the popping of our ears when we sit in a plane, or just hearing the sound of laughter; air pressure is a central part of our lives, whether we are aware of it or not.

But What Is Air Pressure, Precisely?

First of all, what is commonly referred to as "air" is actually a mixture of several elements, all in their gaseous state. The major part of it, nearly four fifths, is nitrogen. More specifically, two nitrogen atoms bonded together. The next common element in the air, about one fifth, is oxygen – also bonded together in pairs. Roughly one percent is the noble gas argon, and about four hundredths of a percent is carbon dioxide. Finally, there is a large host of other gases which appear only in traces.

Crashes in the Air:
Molecules on Speed

All of these air molecules don't just float – they whizz around with enormous speed. Nearly a thousand

miles per hour is what an air molecule has on its odometer, on average! But they cannot enjoy a quiet ride for very long – the air molecules around us can only fly about a couple of nanometers, before they bump into another one. Well, these little buggers are sturdy, and such a collision doesn't break them. It just changes their direction, which is why the air around us can be thought of as the world's largest pinball machine – with billions of trillions of tiny playing balls!

So even on a lovely day, when the air seems to stand still, what is actually going on is a tumultuous hustle, a shoving and kicking of uncounted, microscopic molecules. And we feel this hustle, the thousands upon thousands of impacts per second – as air pressure.

So air is actually a quite dynamic substance, and this is why air appears to try to even out any differences in air pressure, whenever possible. If there is a lower air pressure somewhere, this means that there must be fewer molecules in that region, compared to everywhere else. Now imagine an analogy: picture a crowded room full of people, continuously bumping into one another. Then think of what would happen if you removed, say, one of the tables in this room. Suddenly, there would be an empty space, without any people in it. But soon, the first person would be pushed into the free space, and then another and another. Without any conscious effort by the people in the room, the empty space would be filled immediately. It's the same with differentials in air pressure: the air molecules don't specifically try to spread out evenly, it just happens automatically because of all the continuous bumping off of one another.

The Air Gets Thinner at the Top

There is one exception for this: the higher above the surface of the Earth you are, the lower the air pressure becomes. Why is that? Well, because of gravity, of course!

To understand this, it's crucial to realize the molecules in the air don't all have the same energy! By constantly bouncing off of one another, the molecules exchange some energy, granted. But here is a central fact of physical systems with many, many individual parts which constantly exchange energy: In the long run, when they are all in equilibrium, they won't all have precisely the same energy. Rather, some of them will have more, and some of them will have less, on average. Because of the constant energy exchange, which particle has more energy than the other will change very quickly. But the fact that some particle will have more energy than the average will always be the same. The same

statement also holds for those which have less energy than the average.

So in fact, the distribution of energy among all the different air molecules will not change over time. There will always be some which have more energy than the others, and it's those particles, which can be also found higher up in the atmosphere. To run up against the pull of gravity, you need quite a lot of energy, and the more you have, the higher up you can go. Very high up, there are only those particles with the most energy, and there are very few of those. This is why the air gets thinner and thinner the higher up you go. For every five miles you are above the ground, the pressure in air drops by roughly three quarters. So there is no height at which the atmosphere just stops – rather, the air gets thinner and thinner and thus turns continuously into vacuum.

"Out of the Airlock with Him!"

Speaking of the vacuum – how dangerous is it, really? If one were to be kicked out of a space ship without any protective clothing, what would happen?

Well, one immediate thought would be that one should explode: The human body is made in such a way that it can withstand the pressure of the atmosphere, by pressing with a similar pressure from the inside, to reach an equilibrium. If the pressure from the outside is suddenly gone, shouldn't that internal pressure inflate, and finally pop a human body?

Actually, the fate of such an unfortunate person would not be quite as grim: The whole weight of the air onto our skin is only a few hundred kilograms. While certainly not very pleasurable, the human body – most importantly our skin and our arteries – are strong and flexible enough to withstand that. There is a part which would be ruptured immediately though: our lungs! Consisting of thousands of very fine bubbles, the difference in pressure would certainly rupture the lung of somebody being kicked out of an airlock. So first rule of surviving as long as possible in space: exhale! Then you might, in fact, survive a bit longer.

It would take a few seconds until your body would have used up all of the oxygen still in it. So, before you would become unconscious, there would still be enough time to realize how your body would expand a bit – but not burst! With no atmospheric pressure to hold them back, gases and fluids would begin to leave your body. Jim Le-Blanc, an astronaut who was accidentally exposed to near vacuum in 1965, reported that he could feel the saliva on his tongue boiling away. It wasn't hot, just going into the gaseous state, as if there was sparkling water in his mouth. By the way: LeBlanc was rescued after a few seconds, after he lost consciousness. He completely recovered, so our body seems to be sturdier than one would expect.

The accounts of several accidents (and, unfortunately, also animal experiments), revealed that the maximum time someone could be exposed to the vacuum and still survive, is about two minutes, one expects. But remember to exhale!

HOW ARE YOUR GASSES AND FLUIDS HOLDING UP?

=BRAPP!!

*"Well of course you have no water pressure
— look at the size of your pipes!"*

Fluid Flow and Turbulences
Nothing You Should Test on a Highway

What are physicists good for? Well, it depends on the field in which they are specialists. Some can make a night romantic by telling you a lot about stars and galaxies. Others might be able to fix your car or your TV. But what they should all have in common is to be able to calculate the movements of objects if they know the forces acting on them. This is something fundamental that they learn during their first mechanics lectures. If you follow the rules of Newtonian mechanics, you can treat objects as something point like and get, as a first approximation, a good result. Just to mention a few practical examples: You let an apple fall down a well. If you stop your stopwatch at the point at which you hear it hitting the ground you can calculate how deep it is. You can also calculate how to defeat an enemy in a judo fight who is bigger and stronger than you. Or you can calculate the perfect angle to hold your garden hose if you want to reach the flowers which are the furthest away from you (45 degrees). Feel free to think of further applications of the 45 degree solution.

Movements of Gases and Liquids

But what if you want to calculate the movement of something what is definitely not a point-like-object? Something that is more like a giant bunch of objects, all being influenced by an external force, and also interacting with each other? Then you are entering the field of gas/liquid flow and turbulences. It can indeed get so complicated that even the best computers cannot provide perfect solutions. But for certain conditions you can still get some simple results and fundamental laws that the dynamics of fluids follow. The term "fluid" refers to both liquids and gases. Let us take a look at a fluid, say water, flowing through a pipe. If the velocity is not too high and nothing disturbs the flow, the fluid will move in straight lines. We call this a "laminar flow". This will look similar to a highway with several lanes, filled with cars that all have the same velocity and are not changing lanes. What will happen if suddenly one lane is blocked? Right, we will get a traffic jam. But the reason is not some fundamental law of nature; it is the behavior of us human beings. We are surprised, we slow down, we don't know what the others will do, we don't want to collide and so on and so forth. Let us see what water would do. Look at a tube that suddenly gets tighter, as we have shown in the illustration. If you assume it is an incompressible fluid, which water is, you can say that whatever gets pushed into the tube has to come out on the other side. This thought leads to a law called "continuity equation": The product of a cross section of a tube and the velocity with which the fluid is moving is constant. This means: the narrower the tube, the faster the water! Back to the highway: if all cars would simply go much faster through the part with the blocked lane, there would be no traffic jam. This is of course impractical and dangerous and shows you that you should always crosscheck a physicist's advice with your experience.

Slow Flow

Fast Flow

Bernoulli's Equation

Next to the velocity, we can bring a liquid's pressure ([↗1]) into play. You might know that if you dive deep underwater, the water pressure will increase. In the 18th century, the physicist Daniel Bernoulli derived an equation which connects the height, pressure and velocity for an incompressible flowing fluid with no viscosity. To a certain extent, you can also apply it to compressible fluids, such as gases. Bernoulli's equation states that

$$p + \rho\, g\, h + 1/2\, \rho\, v^2 = constant$$

p is the pressure, ρ the density, g the gravitational acceleration, h the height and v the fluid's velocity. The sum of these three terms will be constant. Let us show you a consequence that might surprise you. You can test it with a little experiment. Place two empty cans on a table and leave a little space between them. Then take a straw tube and blow some air parallel to the cans, the way that Erwin does it. What will happen? Intuitively, one might think that the cans will start repelling each other. Test it.

Did you? The cans started attracting each other. And Bernoulli's equation shows why: Increasing the velocity of the air between the cans has to lower the pressure, as the sum of all terms has to remain constant. If the pressure between the cans is lower than outside the cans, this causes a forces which makes the cans coming closer. Bernoulli's equation plays an essential role in the answer to the question "Why

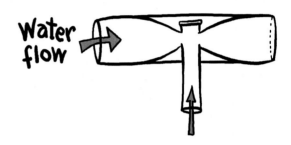

does a plane fly?" ([↗2]). Another nice application is the Venturi tube. The tube has a more narrow part in which the liquid has to flow faster and causes a pressure drop. If you connect a second tube to that part, you can inject a second fluid via the depression, for instance. Such Venturi tubes are used to mix gasoline and air within a combustion engine or to add air to wine to make it taste better.

Friction, Viscosity and Turbulences

Often fluids are not as ideal as we have considered them so far. There is indeed a friction within the fluids which leads to a resistance against a flow. It can be quantified as the "viscosity" of a liquid and tells us "how thick a liquid is". Honey, for instance, has a large viscosity while water does not. You can read more about it in the chapter about superfluids ([↗3]).

[↗1]: *"Vacuum and Air Pressure"* on page 23
[↗2]: *"Why Does a Plane Fly?"* on page 31
[↗3]: *"Superfluidity"* on page 183

The flow of a liquid does not only have to deal with the internal friction of a liquid, but also with the friction of anything that stands in the fluids way. Think of a car driving fast on the highway. The air will flow around the car with a certain velocity. Due to the friction between the air and the car, the car experiences a certain force. This drag force increases with the velocity of the air (or the car) squared. Twice the speed, four times the resistance! And that force can be quite annoying and leads to a higher consumption of fuel. If you want to minimize it, you can either go slow or, probably more convenient, optimize the shape of your car. The shape determines the factor by which

the drag force increases with the velocity squared: the drag coefficient c. Companies put a lot of effort into optimizations of the designs of cars, trains and other vehicles. You can vary your own shape when the next strong wind blows and test the drag coefficients that we provided.

So far we considered the flows of liquids as laminar. But if the velocities of fluids get too high, laminar flows become turbulent. This means that the fluid's pressure and flow velocity changes rapidly with time and space. The moment at which the transition between laminar and turbulent motions occurs depends on the fluids viscosity. The larger the viscosity, the higher the velocity can be without causing turbulences. Turbulences are also called "chaotic". This means that small changes in the conditions of a system cause large changes in its behavior. Think of a river with a laminar water flow into which you put your finger. The water will keep flowing in a laminar way around it. And now think instead of a turbulent flow, such as the flow of milk that you add to your coffee.

You can try to pour milk into a cup of coffee the same way several times, but the swirling patterns will always be different. That is what physicists call chaotic. A general description of fluid motions can be calculated via the "Navier–Stokes equations", which are quite complicated. They contain all the special cases that we have mentioned in the text. On the one hand, they are quite powerful. On the other hand, they are very hard to solve. Calculating turbulent fluid motions requires large computational efforts. If it was not that complicated, the weather forecasts would also be more reliable.

Shape — Drag Coefficient

Sphere → ◯ 0.47

Cone → ◁ 0.50

Cube → ☐ 1.05

Streamlined body → ⬭ 0.04

Measured Drag Coefficients

Why Does a Plane Fly?
How to Guide the Air to Keep You up

If we pick something, hold it up in the air and release it from our hands, it will usually fall. Some things, such as birds, do not fall. This is because they know how to deal with the surrounding air in a way that they can stay up high, or even go higher. Watching a flying bird can be beautiful, but watching a Boeing 737-900 airplane with a loaded weight of up to 79 tons is certainly impressive, isn't it? Have you ever wondered why that is possible? Let us tell you!

One principle that is used to make objects fly is Archimedes' principle: if your density is lower than the fluid in which you are floating, you will experience an upward force. So if you have a balloon and fill it with a gas that is lighter than air, that balloon will rise upwards, if you don't hold it tight. Helium filled balloons work that way. Hot air balloons rise as well, even though they are simply filled with air. But that air is, the name says it, hot. And a hot gas is less dense than a cold gas. But airplanes are neither filled with helium nor with hot air. So what keeps them up in the air?

Action and Reaction

It is useful to ask about how to "keep" it up. We know that a train will stand still if you turn the engine off. A plane will not. It will fall down to the ground. To stay up, it definitely has to move. So somehow the secret has to be related to the plane's velocity. Let us start from the beginning of a plane's flight, down on the ground. Unlike a car, an airplane has no motor that makes its wheels turn to set it in motion ("electric taxiing" is used more and more,

where little taxis with electro motors move the plane to runway, saving fuel). Instead, it uses its turbines (alternatively, you can also use propellers or rockets) to move forward, even on the ground. Air is pushed out of the turbine with high velocity. According to Newton's laws and the conservation of momentum (↗[1]), such a force has to come along with a reacting force in opposite direction. Pushing the air out the back hence leads to a forward thrust of the plane. And here we go: the plane moves. If we take the point of view of the plane, it is standing still, but the air is moving around the plane instead. The details

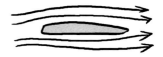

SYMMETRIC WING SHAPE:
SYMMETRIC AIR FLOW,
AND NO LIFTING FORCE.

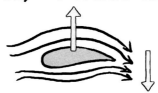

ASYMMETRIC WING SHAPE:
DOWNWARD FORCE
OF AIR, AND UPWARD LIFT.

↗[1]: *"Conservation Laws" on page 51*

of the flow of the surrounding air around the plane's wings is crucial to understand why the plane does not only feel a thrust, but also a lifting force that brings it up in the air. Let us take a look at the shapes of the wings. Each wing is asymmetric in shape. The special asymmetric shape leads to a downward deflection of the air streams. No deflection without a force! And the same way that the counterforce of the air from the turbines leads to a thrust of the plane in the forward direction, the counterforce of the deflected air leads to a lift of the plane!

While the shape of the wing already leads to a downward deflection of the air and hence to a lift, you can increase the effect by tilting the wing with a certain "angle of attack". Using this angle of attack – and in particular its modification during a flight – also allows an airplane to fly on its back. Here, the shape of the wing would, without an angle of attack, lead to an opposite effect and the plane would fall down. By the way: the "more tilt leads to more lift" principle only works to a certain extent. If you overdue it, the air stream will detach from the wing, the lift will be lost and the plane crashes.

Bernoulli's Lift

Next to the downward deflection of the air streams, another effect plays an important role. The wing's special shape leads to a higher velocity of the air stream on the top than on the bottom. If you remember Bernoulli's law from the chapter about fluid flows and turbulences (↗²) you know that faster air

streams lead to a lower pressure. The difference in pressure between the top and the bottom of a wing also adds a lifting force.

So you take the sum of the Bernoulli effect (larger velocity on top than on the bottom) plus the downward deflection of the air, plus some other effects that go beyond the scope of this book, and get a total lifting force. The higher the plane's forward velocity, the larger the lifting force. While the plane is still on the ground, the thrust must increase until a speed is reached that makes the lifting force higher than the gravitational force that pulls the plane to the ground. Once you reach the desired altitude you can set the velocity and the wing configuration such that the lifting forces cancels exactly the gravitational force.

Sometimes you might experience some turbulences during your flight. This happens when suddenly unexpected forces act on your plane. This might be strong winds blowing or thermals, streams of warm air that rise from lower to higher altitudes of Earth's atmosphere.

Not to Forget: Navigating!

So now we know why and how your plane can make it up in the air. But surely you would also like to navigate to a certain place and then come back down to earth safely. Let us tell you the basic options that you have to manipulate the movement of

↗²: *"Fluid Flow and Turbulences" on page 27*

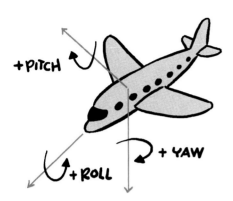

down changes the plane's total angle of attack against the air stream. So if you move them away from their default position parallel to the horizontal plane you induce a force that gives the tail a kick up (if you lower the elevators) or down (if you raise them). This will then cause a pitch by lowering or raising the plane's nose and the plane will go down or up. Attached to the main wings are the "ailerons", little flaps on the wing's back that always move in opposite directions. Moving them increases the lift of one wing and at the same time decreases it for the other. This will cause the roll.

a plane. The best way to describe plane maneuvers is via rotations around certain axes. You can see these three axes in the following sketch. Their names refer to the possible rotations. The "roll axis" points from the tail to the nose of the plane. The "pitch axis" goes from one wingtip to the other. And the last axis, the "yaw axis" point upwards from the plane's center of gravity, perpendicular to the wings. In order to cause the roll, pitch and yaw movements we need some more movable parts of the plane.

The "elevators" are little horizontal flaps, usually located at the plane's back. Moving them up and

A third type of flap, the "rudder" is usually vertically attached to the plane's fin at its back. If you move the rudder to the left or right, the air passing the fin will push against the rudder, leading to a force that causes a yaw. This yaw is then simply a move to left or right, without rolling.

You can also combine several moves: a fast turn of the plane can be achieved by using the ailerons and the rudder the add a roll and a yaw. There are some nice animations that you can find on the web that will certainly help to understand these movements in detail (↗³).

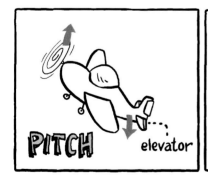

↗³: https://howthingsfly.si.edu/flight-dynamics/roll-pitch-and-yaw

"I don't care if it means a Nobel for maximizing surface tension — I want my pool back!"

Surface Tension
Minimal Surface for Maximum Comfort

Everyone knows that water – at least at room temperatures and normal pressure – is a liquid. That means that it cannot keep a solid form. Rather, it will flow to fill any space that you allow it to. It offers absolutely no resistance to have its shape altered.

This is so obvious to us that the non-solidity of water (or any liquid, for that matter) seems to be completely natural. But how is it, then, that some insects can actually walk over a pond without sinking? Why is a paper clip, carefully placed on a surface of water, able to float? It is made of metal, so why doesn't it sink? And finally, why can one fill a glass of water with more water than actually fits inside?

The answer is: the surface of water is not completely malleable. This is different from the water itself under the surface, which poses nearly no resistance to being deformed. The surface, on the other hand, has a certain stiffness, which is called surface tension.

To understand surface tension, one needs to have a closer look at the way in which the different water molecules arrange themselves, when water is in its liquid form. You all probably know that water molecules consist of one oxygen (O) and two hydrogen (H) atoms, which stick together. But just because the hydrogen atoms in one molecule are attached to their oxygen atom, doesn't mean that they won't try to get close to different oxygen atoms in other water molecules as well. Hydrogen is quite … unfaithful that way. And the oxygen doesn't seem to mind: it gets some connection from further hydrogen atoms in other water molecules as well. These connections between hydrogen and oxygen atoms from different molecules are also called hydrogen bonds.

It's Just a Phase:
the Different Forms of Water

The key point here is that water molecules are not just all by themselves, but there is a slight attraction between different water molecules. This attraction is not very strong, but it is there, and it plays a very important role in what phase the water is in. There are basically three main possibilities:

If the water is very cold, then the molecules are all very slow, and don't have a lot of energy to move about much. Then their mutual attraction is strong enough to keep them close together. In that case, water is solid. That is also called ice (the precise way in which the molecules stick together, in order to form those beautiful ice crystals or snow flakes, is a matter for a different discussion).

If the water, on the other hand, is very hot, then the molecules are very fast, and have enough energy to fly around. In that case, the attractive force from the hydrogen bonds is by far not enough to keep them together. The water takes on the form of a gas, where the individual atoms all zip about individually. Occasionally, they bounce off other water molecules when they hit each other, but apart from that, each molecule stays by itself.

But there is an intermediate case, when water is not too hot, and not too cold: In this special case, a water molecule has enough energy to not be completely stuck to the other molecules. Rather, the hydrogen bonds can occasionally be relaxed a little. But a molecule does not have enough energy to be completely free of all its neighbors. This is the crucial point: it always needs some other molecules to be in its near vicinity. But these don't necessarily always have to be the same individually. Rather, some molecules can leave, as long as some others take their place. So the molecules can shift their positions compared to another, and this is what makes water so completely malleable.

Everybody Wants to Be on the Dance Floor

While solid ice is like lots of people in a packed subway where nobody can move, and water vapor is similar to a wide open field where a few people run around, water in its liquid phase can best be compared to people on a dance floor: usually they are reasonably densely packed with people, but not so much that the dancers cannot move and look for new dancing partners if they want. And those people at the outer edges of the dance floor always try to get further inside, because dancing is best when you are surrounded by people!

This is pretty much the situation with water molecules – and in fact nearly every other liquid. The ones on the inside are surrounded on all sides by other molecules, with which there is a slight attraction. These particles can move rather freely everywhere they want, as long as they are always surrounded by other molecules. But, the ones on the water surface only have other attractive molecules towards the inside – the air molecules on the outside on the other hand are not attractive to them at all. So they crowd together and try to get inside. This puts the whole surface under a tension, which makes it stiffer than the inside of the liquid. In fact, it makes it stiff enough so that smaller insects can walk on the water surface without sinking.

Size Does Matter: The Smaller the Surface, the Better!

This surface tension has a very important consequence for the shape of water: If there are no other forces acting on it (like gravity, or the pressure from a surf board, or tidal forces), water always tries to get into a shape so that it has a surface area that is as small as possible. For a given volume, that is the sphere – which is why water forms droplets.

So if an astronaut in the International Space Station spills her drink, for instance, it will not splash anywhere, but rather float around in lots of small droplets. Here on earth though, gravity is much stronger than the surface tension of water. So if you fill it into a funnily-shaped glass, gravity will squish it to fit into the shape. But for instance, the situation is slightly different with mercury: This is a liquid metal (at room temperature), and its surface tension is roughly seven times higher than that of water. If you spill a glass of mercury, it will quickly form into small spheres that lie around on the floor. Please, never actually do this – mercury is really quite poisonous!

Ordinary leaf

lotus leaf

is because, if the water actually were to be in contact with the leaf everywhere, its surface would have to be very large – it would have to fill out every valley in the leaf's topography. This is what the water tries to avoid, so it will pull itself together to be a droplet. That way, it touches only very little of the actual leaf – and instead of slowly dripping off, it can just easily "roll off". On its way, it takes quite a lot of dirt with it, rather than floating over it. So the Lotus plant uses the surface tension of water in order to keep itself clean!

RELAX, YOU'RE TOO TENSE!

Speaking of cleanliness: One reason for using detergent in a washing machine is that soap and other substances that contain surfactants lower the surface tension. So a water strider cannot walk on soapy water! And such water is much more able to get into all the small crevices and folds of the clothing in the washing machine – for which it needs a large surface.

Cleanliness Is Next to Godliness – with Surface Tension!

The fact that water always tries to have as little surface as possible, is used by the Lotus plant in order to protect itself from dirt: Most plant leaves have a rather smooth surface. Rain water tends to stick to their leaves, and keep them wet. The leaves of the Lotus plant, however, have a very peculiar surface structure: under a microscope one can see that the surface is covered with lots of little hills, as if the leaf had goose-bumps. Now, if a water droplet sits on the leaf, it will not touch the whole surface, but rather only the tips of the little goose-bumps. That

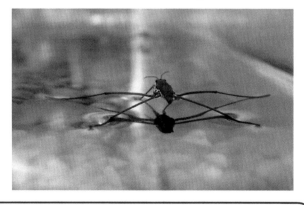

Image: Tim Vickers

Non-Newtonian Fluid
Is It Liquid? Or Is It Solid?

How could you characterize a liquid? You could start with its color, maybe talk about its smell or taste (watch out!) and consider its acidity (lime juice or soap). But sooner or later you will think about the liquid's viscosity. The viscosity quantifies what we use to call a "thick" or "thin" liquid. High viscosities lead to thick liquids (such as honey), low ones to something like water. You can read more about viscosity in the chapter about superfluidity (↗[1]), where we introduce a liquid that has no viscosity at all. It is more liquid than water and will not even stay in a cup into which you put it in. Instead, it will crawl out.

But back to liquids and their viscosity in general. We say that a liquid has a certain viscosity. Instead of liquid we can also talk about fluids, which includes gases and liquids. Now, can you imagine that the viscosity of a fluid can change with the fluid's velocity, stress or pressure? If it does not, we call it a "Newtonian fluid". But if its viscosity does indeed change, it is called a "non-Newtonian liquid". The change in viscosity by the application of a so-called shear stress can be either positive or negative, so the viscosity can increase or decrease. Forces that cause a relative movement between different layers of an object cause shear stress. Sounds complicated, but it is simply what happens if for example a fluid is pushed through a tube and feels a friction on the tube's wall.

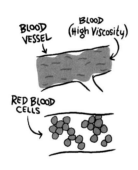

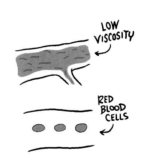

So wherever a fluid is moving you will get a shear stress. And while this usually does not change the fluid's viscosity, we now take a closer look at liquids where it does.

Where Pressure Helps – Shear Thinning Fluids

We hope that you've never had an uncomfortable situation in which you had to touch too much blood. But maybe you already are aware that it feels more viscous than water. As the blood vessels can be pretty thin, you might wonder how the thick blood makes its way through. The reason is that blood is a good example for a non-Newtonian fluid. It belongs to the group of "pseudo-plastic" or "shear thinning" fluids. This means that for increasing shear stress the viscosity decreases. Very good! So flowing blood gets less viscous and can even make it into the smallest vessels. The illustration shows the effect. Blood is more than a normal liquid; it contains a lot of cells. The erythrocytes, known as red blood cells, are mainly responsible for the non-Newtonian behavior. In large blood vessels without a lot of pressure, they tend to form clusters that block each other during movement. In thinner vessels, the stress does not only break these clusters but can also deform the red blood cells. The

↗[1]: *"Superfluidity" on page 183*

optimization of their shape leads to a better flow and a lower viscosity.

Another famous example for a shear thinning fluid is ketchup. Its large viscosity makes it hard to get it out of the bottle. But the application of a shear force via bottle shaking makes it less viscous, and the ketchup can get out. Sometimes the effect is even larger than desired and instead of fries with ketchup you end up with ketchup with fries.

Shear Thickening Fluids – Hitting Hurts

The more curious of you might not just want to read about it, but also to test it. The good news: we will now introduce a fluid which is non-Newtonian, is great fun to play with and can be self-made easily. It belongs to the group on shear thickening fluids, so it basically has opposite behavior with respect to blood and ketchup. If you apply stress, its viscosity will increase quite dramatically. The fluid we are talking about is what Erwin is having for soup. You might know it as "Oobleck". The name comes from the children's book "Bartholomew and the Oobleck", written in 1949 by Dr. Seuss. It tells the story of the young boy Bartholomew, living in the kingdom of Didd. Didd's king got bored by the rain and wanted something different to fall from the sky. And so it happened: a sticky substance called Oobleck covered the whole kingdom and Bartholomew had to rescue it.

The Oobleck that we are talking about (not the one from the book) can be produced easily: All you need to make a batch is some water and cornstarch. Mix

it in a ratio of 1:1.5–2. Once you know what is supposed to happen you can vary the mixing ratio until you get a perfect result. You should slowly mix the

two until you get a substance that looks like a liquid at first sight. You can then try to make a quick movement of your finger in the Oobleck. Get it out quickly and it will stick to your finger! Try putting your finger in quickly and it will block you – just like a solid.

A full and complete explanation of the shear thickening effect of Oobleck does not yet exist. The "getting solid" effect is also said to be more than just an increasing viscosity. But one common explanation has to do with the relatively large cornstarch particles. Surrounded by water molecules, they can smoothly move. But once the external force is applied, the water gets pushed out of the space between the cornstarch molecules, causing a strong friction between them that makes the Oobleck feel sticky and solid.

The more Oobleck, the more experimental possibilities! If you fill a bowl you can try punching it. But watch out, you might hurt yourself. You can also take a bit of Oobleck in your hand and move it quickly. The Oobleck will stay solid and you can carry it with your hand. But as soon as you stop the movement, it will run out of your hand (watch out, you might create a mess!). Do you have a loudspeaker? A big one? One that is maybe not brand new and superexpensive? Then you can remove its cover, put some plastic wrap over it and place some Oobleck there. When you turn on some music, the Oobleck will bounce and create funny solid structures while bouncing.

The queen of Oobleck experiments needs a big pool, completely filled with it. And it requires a brave experimentalist. If you're quick on your feet you can walk across without sinking! But don't stand still on your way, as this will make you sink. You can find some very nice videos on the web if you look for Oobleck. But nothing is as good as testing it yourself, right? You can also make it look nicer if you add some food coloring. Enjoy! Oh, one word of warning: after some time, the water and cornstarch will separate. So don't put the used Oobleck down the drain as it might clog your pipes. Use the trash instead.

While experiments with Oobleck are definitely entertaining, people also think of applications of substances with such a non-Newtonian behavior. Putting such a liquid into a vest would feel comfy. But as soon as something hits you hard (like a bullet) it can become rock-solid and absorb the impact.

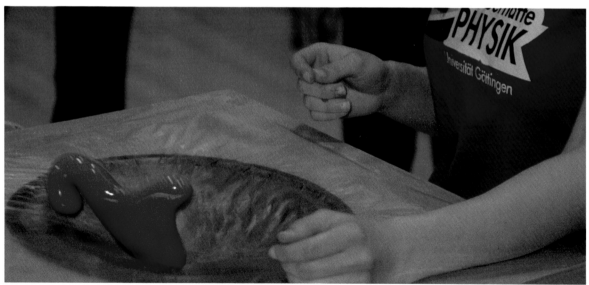

Image: Arnulf Quadt / Georg-August-Universität Göttingen

"Nice maneuver, but you can't leave it here."

Rocket Maneuvers
Navigating within Nothing

Satellites are doing a great job. They orbit Earth and help us to establish worldwide communication, bring us TV signals, help us to navigate, look at the weather, produce maps of Earth and many other things. Did you ever think about having your own satellite? To avoid all the complications of developing the satellite itself, let us just think about bringing it up in space. To simplify the problem even further, let us think about getting a tennis ball with our name on it into space. The simple approach: Take one and throw it up in the air. What will you observe?

Too Fast to Stay

Most likely, the ball will fall back down to the ground. But you will also observe that the faster you throw it, the higher it will go. The reason why it comes back is the gravitational force that acts between Earth and the ball. This force is the reason why we are all "attached" to the surface of Earth and can walk on it without getting dragged into space. One can say that gravitation binds us to the earth. This binding can be quantified with a certain energy. If you throw up your tennis ball, it also has a certain amount of kinetic energy. The faster you throw it, the more it gets. This energy cancels some part of the binding energy of the gravitation. At the turning point of the ball, it has lost all its kinetic energy and regains it by being accelerated back to the ground. Doing a bit of calculus allows you to get a solution to the question: "So how fast do I have to throw my tennis ball to

give it more energy than it has gravitational energy?" This velocity – called "escape velocity" – is your ticket into outer space. It depends on Earth's mass and its radius. It has a value of 11.2 km/s, 40,000 km/h or 25,000 mph. This is pretty fast. For the sun, the escape velocity is much higher: 617.5 km/s.

This escape velocity is indeed the minimal velocity that you need to have in order to leave Earth and go into outer space – but this applies only in the case that you do not have any further propulsion, as in the example of a thrown ball: once it left your hand, its velocity will decrease. But rockets can keep accelerating and that's what allows them to start with much lower velocities. An example for an object without propulsion for which the escape velocity is more important are atoms. Hydrogen and helium are pretty light atoms. The temperature of the gases of our atmosphere determine the average kinetic energy of their atoms. For an equal temperature, and hence an equal kinetic energy, light atoms are faster. This is the reason why our atmosphere does not contain any hydrogen or helium: unlike for the heavy atoms like nitrogen and oxygen, the hydrogen's and helium's temperature leads to velocities which are larger than the escape velocity. That's why they leave Earth.

MULTISTAGE ROCKET

TANK 1
HALF EMPTY

TANK 2
FULL

EJECTED

TANK EMPTY

By the way: You can decrease the escape velocity if you make use of the fact that Earth is spinning. Depending on the direction of your start (against or in direction of Earth's rotation) you can increase or decrease the escape velocity by about 10%. As Earth's rotational velocity is largest at the equator, many space launch facilities are located in this region, such as for example the American Cape Canaveral or the European Guiana Space Center.

Go Rocket, Go!

We know now that each journey into outer space starts with escaping the gravitational field of Earth. The question is: how can we reach that speed? Typically we use rockets. They follow the principle of ejecting

high-speed jets of rocket propellant. According to Newton's laws this leads to a thrust of the rocket.

You can keep accelerating a rocket as long as there is still some propellant left. The interesting point about rocket propulsion: it has to accelerate not only the mass of the rocket, but also its own mass (at least before it leaves the rocket). So the more propellant is used, and expelled, the easier it is to accelerate the rocket. This fact is considered in the famous "rocket equation". It tells us that the maximum velocity that a rocket can reach is determined by the velocity with which the propellant leaves the rocket multiplied with the logarithm of the ratio of the masses of a full and an empty rocket.

So for maximum rocket speed you can either try to maximize the exhaust velocity of the propellant or the ratio of a full to an empty rocket. The rocket equation also tells you that it is more efficient to use a multistage rocket instead of a single large rocket. Multistage rockets have several stages, each equipped with engines and propellant. After each one is empty, it is detached from the rest of the rocket. Next to the optimization of the mass ratios and the usage of multistage rockets you can also increase the rocket's final velocity by increasing the exhaust velocity of the propellant. Classical rocket propellants exist in solid and liquid form and are used in combustion engines. The good thing about that: they even work in a vacuum, as in outer space. There, a classical airplane's turbines could not work without any surrounding air. Oh, one thing: the rocket equation that we quoted did not take any external forces into account. If you build a rocket car (a car with the rocket attached on the roof, making it really fast) with negligible air resistance, that's fine. It will only move horizontally and can ignore gravity. But for a rocket, you should better take into account the effect of gravity. Nevertheless, the statements about the dependence on the full to empty mass ratio and the exhaust velocity remain valid. For rocket movements out in space, gravity can be neglected.

Maneuvers in Outer Space

Once you made it into outer space you can perform several maneuvers, next to a simple "staying in an orbit" or "flying in a direction". Imagine you are surrounding Earth, but want to change your distance to Earth by changing the orbit. An interesting one is, for example, the geostationary orbit at about 36,000 km above Earth's equator. Remember that on each orbit you need a specific time to circle the Earth (or any other object in outer space, ↗1). In a geostationary orbit this time corresponds to the time that earth needs to rotate. This means that a satellite in a geostationary orbit will have a fixed position from Earth's point of view. A maneuver that is used to change the orbit is called "Hohmann transfer". If you want to do a Hohmann transfer to move further away, you need two little impulses, each in the direction of flight and tangential to the orbit. The first

one will bring you into an elliptic path, the actual Hohmann orbit. The second impulse of the same type as the first moves our spacecraft out of the Hohmann orbit into a second, circular path. Hohmann transfers of course also work the other way around by reversing the impulses.

Another maneuver which sounds romantic but is quite complicated, is a "space rendezvous": two spacecrafts meet in the same orbit at the same place. One of it is passive and waits; the other one is active and approaches the passive one. It sounds quite easy: first you do a Hohmann transfer and get into the right orbit. Then you just have to get closer to the passive craft. This you can do by applying a little thrust. But wait! This will change your velocity which will lead to a change of orbit. So it involves a little more thinking, orbit changing and impulses.

Last but not least let us introduce the "swing-by maneuvers". If you plan a space mission to an object far away, let's say to Mars, you want to save every bit of propellant that you can. Swing-by maneuvers accelerate your spacecraft via the gravitational force of other planets which are met on the way and fly in the same direction. You can even use multiple swing-by maneuvers as it was done for the Rosetta space probe launched in 2004 (↗2). On its way to the comet 67P/Churyumov–Gerasimenko it performed three swing-bys of Earth and one of Mars. Such a sophisticated journey requires perfect timing! "Oops, we missed Mars" would have meant the end of the mission.

↗1: *"Kepler's Laws" on page 47*
↗2: *"The Voyager Probes" on page 55*

"They say they're from a planet that broke Kepler's Laws."

Kepler's Laws
The Basic Rules for the Movement of Planets

Imagine you are a popular scientist (maybe you are) and you are dealing with a hot topic. Suddenly you realize that there is someone else, very talented, not yet that popular and working in the same field as you. What would you do? Collaborate with him? Or put obstacles in his way? Or something in between? One of the most important astronomers, the Danish Tycho Brahe, was exactly in such a situation. But let us start from the beginning.

Brahe observed the movement of stars and planets at a time where the telescope was not yet invented, namely at the end of the 16th century. He did a brilliant job without telescopes, which were developed just a few years after his death. One of the big challenges that he faced was the explanation of the planetary orbits. At that time, people still believed that the Earth was the center of the universe, and that all of the planets, and the sun, revolved around us.

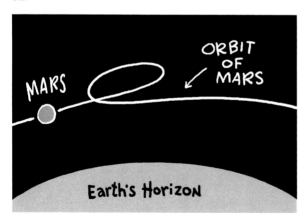

But that model failed to explain some strange details of planetary orbits, for example the one of Mars. You can see the observed trajectory of Mars in the sky, as seen from Earth in the illustration. It seems as if Mars turns back for a while, until it continues on the line it started. Using circular orbits around the Earth, this observation would not make much sense. People tried to improve the model by putting another little circle on the orbit of the "big orbit" of Mars, around which the planet was supposed to move. But even this idea was not fully convincing.

Kepler, the Young Talent

Brahe had the feeling that he could use some help, even though he had probably not admitted it. He knew Johannes Kepler, a young and talented astronomer. While Brahe had a lot of very precise data from astronomic observations, Kepler had talent to build theoretical models. Getting back to our introduction: If you were Brahe, what would you do? Brahe decided for the last option. He invited Kepler to Prague, the place where he did his research, and asked him to become his assistant. This way he could profit from Keplers talent. But by giving him access to his data, how could he make sure that Kepler would not analyze it quickly, establish new theories and be in the end even more famous than Brahe himself? Brahe decided to keep Kepler busy for a while by giving him a task that seemed to be impossible to solve, at least within a short amount of time. Kepler was made to analyze the orbit of Mars.

This was indeed a very challenging task. Kepler started to solve it with a completely novel approach. In contrast to Brahe, he put the sun in the center of our solar system, not the Earth. We call this system "Heliocentrism" or Copernican system, named after the astronomer Nicolaus Copernicus, who invented it about 100 years before Kepler worked on the Mars orbit.

Kepler's Laws

Even with this system there was no easy way to explain the orbit of Mars. What was still wrong with this model was the idea that the planets move in circles around the sun. Kepler finally discovered that the orbits were not circles, but ellipses! And there was it, Kepler's first law:

The orbit of a planet around the sun is an ellipse, and the sun lies in one of the ellipses focal points.

But what is a focal point? Let us start with the definition of an ellipse. "It is something like a circle, just squeezed or stretched" – this is probably what you have in mind. But the mathematically exact defini-

tion of an ellipse is different: An ellipse is a curve around two focal points for which all points of the curve have a constant sum of distances to the two focal points. Wow, this is tough to understand. Think of two wooden sticks in the ground (our focal points). Take a string that is longer than the distance of the two and attach it to the sticks. Now take a pen, stretch the string with it and draw a line. As the strings length is constant, all points on the line will always have the same sum of distances to the two focal points: the length of the string. It surely helps to see an example of an ellipse in our illustration. A circle is, by the way, just the special case of an ellipse where both focal points are the same, namely the center of the circle. While those people who believed in purely circular orbits also claimed that the orbital speed was constant, Kepler had to admit that this was no longer the case for elliptic orbits. But instead, another quantity which is related to the planet's movement is constant. And here it comes, Kepler's second law:

For equal time intervals, the line segments connecting the planet and the sun sweep an equal area.

The closer a planet is to the sun, the faster it gets. This is caused by the conservation of angular momentum (↗¹). Think of the figure skaters bringing their arms closer to their body to spin faster. So, close to the sun our planet moves faster on its orbit and also drags the connection line faster. But as it is so close, the covered area will be smaller (per bit of movement). And the faster movement exactly cancels the slower coverage by the shorter distance

SEMI-MAJOR AXIS

empty focal point

SUN

↗¹: *"Conservation Laws" on page 51*

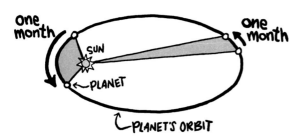

to the sun. Amazing! We have to be aware that all these laws were formed empirically. This means that Kepler only had a set of points, namely the location of planets, and the corresponding times. And he could not see the whole picture from far away, where the orbits can be clearly seen, but from down on earth. So from the strange orbits like the one from Mars that we have seen in the illustration, he derived these two laws. While Kepler had formulated these two laws pretty quickly, it took him some more time to derive the third law. And this third law states:

The square of the orbital period divided by the cube of the semi-major axis of the orbit is constant.

The major semi-axis is indicated in our illustration of an ellipse. It is the average of the longest and the shortest distance between a planet and the sun. Roughly speaking, Newton's third law says: the larger the orbit, the longer it takes to complete it. While this is quite obvious, it also tells you exactly how much longer you will need to orbit the sun. Once you know this constant value, for example from the Earth's orbit, you can calculate orbits for all other planets! Let us try it with an example

case. The mean distance between Earth and the sun is one astronomic unit (1 AU), corresponding to 93 million miles. If we know that Jupiter's average distance to the sun is 5.2 AU, do we then also know how long it takes it to complete its orbit? Yes, with the help of Kepler's third law we do! We connect the ratio of squared periods and set it equal to the ratio of cubed distances. Resolving this equation for the period of Jupiter leads to 11.9 years.

What it so impressive about Kepler's laws is that they were found without even knowing the principle of the gravitational force, which Isaac Newton formulated about 70 years later. It is a nice example of how natural laws are discovered: first, you observe nature. Then you try to find a mathematical rule that nature follows. And in the end you try to generalize and validate that law. Today, as we know about Newton's gravity, we could also start the other way around: take Newton's law of gravity and derive Kepler's laws from them.

Kepler's laws are not only valid for the movements of planets around the sun. You can also use them so calculate the orbits of satellites, for instance, the moon's orbit around Earth. You just have to make sure that the approximation that the two objects are massive and do not get distracted by other massive objects (which is true for the solar system where only the sun is really massive and other objects are quite far apart from each other) and that you can neglect non-gravitational distractions in the orbits. Something like that could be caused by, let's say, too many asteroids in the way of a planet's orbit. But fortunately, this is not the case.

Conservation Laws
Nothing Gets Lost in Nature

Running a power plant consumes resources which we only have in a limited amount (at least for nuclear, gas, coal and oil based ones). You see: there is no reason to waste something. But what actually happens to the energy that we waste? Is it gone? You might say: "Yes, because there is no way to get it back!" But a physicist would probably answer you: "Energy is conserved in nature! Always and everywhere!" And he is right. Our laws of nature are based on the conservation of certain quantities, and energy is one of it.

Symmetries Preserve Things

It is certainly good to know which quantities are conserved in nature. But it is also interesting to know why, isn't it? One way is to observe a lot of processes, pick a quantity and always compare "before" and "after". This is a typical approach for an experimental physicist. But also the theorists have an important tool to actually predict quantities which will be conserved. The female mathematician Emmy Noether postulated a theorem ("Noether's theorem"), proven by her in 1915, which states that "each continuous symmetry comes along with a conserved quantity". What does that mean? Whenever we see a certain symmetry in nature, we can check for the corresponding conserved quantity. Such symmetries refer to quantities which can be transformed without changing the properties of a system. Okay, now it's getting too technical. So let us start with an example. If you throw an apple, you can calculate the way it will fly and the time it takes until it reaches the ground. What if you now trans-

form the time in the equations to a later time? At night, the apple will fly exactly the same way.

You can also think of another transformation: walk a few steps to the side and then throw the apple. Still the same, right? And if you like you can also rotate to a certain angle before you throw it. Again, it will fly the same way (to the other direction, of course). There is no doubt that you can make the proposed transformations without changing the physics in nature, right? And now think of Noether's theorem: with each of these symmetries (invariances under transformations) a conserved quantity comes along! The first one, time invariance, leads to the conser-

vation of energy. The second one, translational invariance, leads to the conservation of momentum. And finally, rotational invariance leads to the conservation of angular momentum.

Energy Conservation – Still There, but Maybe Less Useful

Let us go back to energy. What happens to it, once you "wasted it"? It depends on the process. Let us look at our cartoon. Erwin climbed up the hill. This cost him energy (maybe he burned some fat), but he gained potential energy as he went further up. This potential energy he can then use to fall off the cliff and convert his potential energy to kinetic energy when getting faster and faster. When he hits the ground, his kinetic energy will most likely be converted into thermal energy. The friction between him and Maxwell's arms cause heat. If he fell onto something else his kinetic energy could also be used to change the structures of either him or the thing he fell onto. But let's not think about that.

If you drive your car, have a kinetic energy and then brake, you also convert it into heat (thermal energy) of the brakes. And even if mysterious things happen, such as those click-heat-pads (the gel pads that suddenly start to get warm once you pressed a clicker inside) suddenly producing heat out of nothing, energy is conserved. The professional name for such pads is "phase-change material". By using an

external trigger (the clicker) you start a change of the material's phase from liquid to solid, and energy is released. This energy was in the chemical structure. As the new structure needs less energy to keep its form, it can release the energy and the pad heats up.

Other interesting appearances of energy conservation are particle decays (↗[1]), nuclear fusion (↗[2]) and nuclear fission (↗[3]). In these processes, energy is also partially converted into/from mass via $E=mc^2$ (↗[4]).

Momentum Conservation – Mind the Recoil!

A momentum is defined as the product of a velocity and a mass. And this quantity is called a "vector". In contrast to a scalar, it does not only have a numerical value, but also a direction. What is conserved is the total momentum of a system. In the chapter about the Standard Model (↗[5]) we see Erwin and Maxwell on skateboards. At the beginning, the total momentum is zero. When Maxwell throws the banana, the banana has a momentum. But to compensate this momentum to the right, something else

must get a momentum to the left: Maxwell himself. If his weight is 50 times the weight of the banana, then his recoil velocity to the left is only 1/50 of the banana's velocity.

↗[1]: *"Particle Decays" on page 221*
↗[2]: *"Nuclear Fusion" on page 175*
↗[3]: *"Radioactive Decay" on page 167*
↗[4]: *"E=mc²" on page 237*
↗[5]: *"Standard Model of Elementary Particles" on page 213*

Momentum conservation is everywhere – and is often forgotten. Think of it when you stand close to a cliff and consider throwing your banana! And also think of it when you shoot a gun. Fast bullets can cause quite some recoil. And even if you jump up in the air, you give the Earth a kick in the other direction! But as the earth's mass is about 10^{23} times your mass, earth's recoil velocity is really negligible.

Spin It around – Angular Momentum

What the momentum is for linear motions is the angular momentum for rotations. If you spin a top, it has no linear momentum, but angular momentum. If a ball rolls down a hill, it has both linear and angular momentum. There are several ways to define an angular momentum. For a simple setup where you have an object rotating around an axis, the magnitude of the angular momentum is the product of the mass, the velocity and the distance to the rotation axis. If the product of these three values has to stay constant, you can observe funny effects.

ARMS OUT: SLOW SPIN

ARMS IN: FAST SPIN

If you watch figure skaters doing a pirouette you will recognize that if they move their arms closer to their body, they will spin faster. If the distance of the spinning objects to the spin axis gets lower, the velocity has to increase. Another impressive example is an experiment called fire tornado (you should find some nice examples on YouTube) which you should

definitely see. In astrophysics, angular momentum plays an important role as well and leads for example to extremely fast rotating neutron stars ($\nearrow$6).

And the Rest

Other quantities are conserved as well. Charge conservation tells you that each electric charge must come from somewhere, electrons for example must have been ionized from an atom (neutral atom becomes negative electron and positive ion). Or in case you want to use energy to produce matter, as it is done at particle accelerators ($\nearrow$7), you will see that for each matter particle the corresponding anti-particle ($\nearrow$8) is produced. As the charges of particles and their anti-particles are opposite, the pairs in total are neural.

Next to the electric charge there are also other charges which are conserved during interactions, such as the strong and the weak charge. You can find out more about them in the chapter about the strong ($\nearrow$9) and the weak interaction ($\nearrow$10). The reason why physicists perform very precise measurements to test conservation laws very carefully is because they are hunting for deviations. Any broken conservation law tells us where we have to tune our laws of nature. A popular example is the "CP symmetry" ($\nearrow$10). Violating it can explain an asymmetry between matter and antimatter without which we would not exist.

$\nearrow$6: *"White Dwarfs and Type Ia Supernovae" on page 87*
$\nearrow$7: *"Particle Accelerators" on page 249*
$\nearrow$8: *"Antimatter" on page 217*

$\nearrow$9: *"The Strong Interaction" on page 229*
$\nearrow$10: *"The Weak Interaction" on page 233*

The Voyager Probes
Where No One Has Gone Before

The planets of our solar system have been known for a long time. Our closest neighbors, like Venus and Mars, can be seen with the naked eye on a clear night. Ancient civilizations knew them, and it is no wonder they carry the name of gods. The ones further away, such as Neptune and Uranus, the ice giants at the outer etches of our solar system, have been an inspiration for scientists and artists alike, throughout the centuries.

To Visit the Ice Giants

The human capability for curiosity seems limitless, so it is no wonder that we've always wanted to visit those planets. See what they are made of, and what is beyond them.

Then, in the sixties and seventies of the 20th century, two factors convened which made it possible to send human-made vessels to the outer reaches of the solar system. Firstly, technology had just advanced enough to send unmanned probes and manned vessels into the nearest vicinity of the Earth. So the machinery was there to make limited spacefaring possible. Soon, the plan was formulated: we need to send probes to the outer reaches of the solar system!

Secondly, a specific alignment of planets took place, which was absolutely necessary to be able to go beyond the inner planets. You see, the further away from the sun you want to send your probe,

the more energy you need. To give it enough energy to reach the outer limit of the solar system, you'd need it to lift off with about 25 miles per second! The amount of fuel needed to do that is enormous – and that would have made the probe much, much heavier, requiring even more fuel. Far too expensive!

However, there is a neat trick to give a probe enough energy: the fly-by maneuver (also called "swing-by", ↗[1]). For this, you just need to give your probe enough initial speed to reach a planet – and you need to aim slightly behind the planet's path. As soon as the probe approaches the other planet, it will feel its gravitational pull. If you calculate your trajectory precisely, the probe will not crash on the planet, but just barely miss it. Passing by very closely, the planet's gravity will fling the probe further out into space with force! Essentially, the probe will steal some of the energy from the planet's movement around the sun – but that is so enormously large as compared to the movement energy of the probe that it will not notice the loss. For the probe, however, this means it has gained new energy to travel further away from the sun.

The Planets Were Right in 1977!

Now, in 1977, the planets of the solar system were in a special alignment which happens only every 175 years. Jupiter, Saturn, Uranus and Neptune

↗[1]: *"Rocket Maneuvers" on page 43*

were in such a constellation that a probe could carry out multiple, consecutive fly-by maneuvers, and get energy from all of those planets!

Thus, in 1977, the two Voyager probes were launched. Voyager 2 was the first one to leave the surface of the Earth, designed to perform fly-by maneuvers by Jupiter, Saturn, Uranus, and Neptune. Its mission was to visit all of these planets over the course of three years, and send detailed images and telemetry data back to Earth. Shortly after its launch, Voyager 1 (which you can see in the image) was launched, on a shorter, faster trajectory. Flying by Jupiter and Saturn, it was decided it should pay a close visit to the Saturn moon Titan, which scientists were very interested in.

Jupiter, Saturn, Uranus, and Neptune

Despite being launched later, Voyager 1 quickly overtook Voyager 2. In late 1979, they reached Jupiter, only a few months apart from each other.

In the Jovian system (meaning the planet Jupiter and its many moons), they made a host of interesting discoveries, most notably that the Jupiter moon Io had active volcanoes, which affected the whole area. We also learned that not only Saturn, but also Jupiter has planetary rings!

In November 1980, Voyager 1 flew by Saturn, making detailed measurements of Titan's atmosphere, which had just been discovered one year earlier by Pioneer 11's sensors. For that, it had to slightly alter its course, however. Although it could perform detailed measurements, the changed course had to include a fly-by maneuver around Titan, which shot it out of the ecliptic, the plane in which all planets orbit around the sun. Since then, Voyager 1 is on a direct trajectory out of the solar system, and will never come close to another planet. In August of 1981, Voyager 2 passed by Saturn, and continued its originally planned path to the other ice giants.

It reached Uranus in 1986, where it detected 10 previously unknown moons. It flew by Neptune in 1989, where it discovered a great, dark spot on its surface, not unlike the great red spot on the surface of Jupiter. The Hubble Space telescope

Image: NASA

has, however, confirmed that the spot has vanished by now.

During the 1990s, the Voyager probes overtook the slower deep-space probes Pioneer 10 and Pioneer 11, making them the human-made objects which are furthest from the sun.

On the 14th of February 1990, Voyager 1 was far enough out that it could take a photograph of the whole solar system – including all planets, the Earth being a pale, blue dot.

Leaving the Solar System

In 2004, Voyager 1 reached the termination shock, the region where the solar radiation hits the interstellar medium. At this region, the temperature outside the probe are a few million degrees – no danger for the probe itself, however: although the particles around the probe have an enormous energy, there are just so very few of them (remember, this is basically vacuum). The termination shock is considered to be the beginning of the boundary of the solar system.

Since 2010, it has been confirmed that Voyager 1 detects no sign of the solar wind any more. It has definitely left the solar system!

The Golden Records

Originally planned to run for only a few years, the Voyager probes have been in continued service for nearly 40 years now. During that time, they have gathered invaluable data about the solar system and its planets. On their travel, they have carried the Voyager Golden Records, phonographic plates

made out of gold-plated copper, so that they stand the test of time as well as possible. These records contain sounds and images of planet Earth and its current civilization:

Sounds of crashing waves, singing birds, music by Beethoven and Bach, as well as analog images containing information about human beings, life on Earth, and where to find our planet (relative to 14 very bright pulsars in the galaxy) are contained on the records. It seems unlikely that an advanced civilization will ever find them, given how vast the universe is.

Still, this cosmic message in a bottle serves as a time capsule, carrying some elements of human civilization outwards into interstellar space. Maybe it will be the only thing remaining of us when the sun will swallow up the Earth in a few billion years.

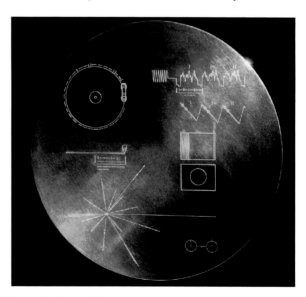

Image: NASA

"Speaking of dust clouds and gas giving birth to things — do you know you've got a mini solar system growing in your washroom?"

Birth of the Solar System
A Star Is Born

The solar system is our home. Its central star, the sun, is a roaring ball of nuclear fire that illuminates eight planets that orbit it, one of which is the Earth where we live. Apart from that, there are several smaller dwarf planets, as well as millions of rocks, asteroids, and smaller chunks of ice, all in revolution around the sun.

Our sun is only one of around a hundred billion stars in our galaxy, the Milky Way. And while it is a bit difficult to say anything precise about it at the time of writing, it appears that most of these stars seem to have planets of their own. So the galaxy, and in fact the whole universe seems to be full of stellar systems. With that in mind, it seems natural to ask, where all these stars and their planets come from. How have they formed?

While none of us was there when the sun started its life, and our own solar system formed, we still have some idea of how that went down. That's because we are in the lucky situation that new stars are born everywhere and all the time in the galaxy, and we can look at how that works with our telescopes.

Another One Bites the Dust:
The Source Material for New Stars

The galaxy is full of dust and gas – mostly hydrogen and helium, but also traces of heavier elements, minerals, and ice – which forms the interstellar medium. This gas is not evenly distributed. Rather, it clumps together in some places more than others. Why? Well, the galaxy is not exactly a quiet place: as one can read in (↗[1]) stars explode at the end of their life, shedding a lot of their outer layers, jettisoning them into space. So there are streaming and swirling clouds of gas everywhere, and sometimes these clouds collide, and tend to stick together, due to their gravity.

At those places where the gas is slightly denser than average, it tends to contract more than average. So wherever there is a surplus of gas and dust, it will attract even more gas and dust, increasing the local density even further, and so on. This is the thing with gravity: it is only an attractive force, never repulsive.

There is no fair spreading of mass in space. Rather, as money makes money, so interstellar medium makes interstellar medium.

So it should not be surprising that the interstellar medium is full of swirling clouds of gas and dust.

↗[1]: *"Supernovae" on page 83*

So there is a huge rotating cloud of gas, held together by its own gravity. It has to be huge, otherwise there would not be enough mass to keep it together. Rather, it would just drift apart. In fact, the size of the cloud out of which our solar system formed, was probably around 20 parsecs, or over sixty light years, in diameter.

Because of its rotation, the cloud flattens like a pancake, becoming thinner at the outer rim, bulging in the middle. Actually, after a while, it looks more like a giant sunny side up than a pancake.

The innermost region of the giant interstellar rotating disc (the yellow part of the egg) contracts under its own gravity, and the atoms come very close to each other. Pressure and temperature become enormous. In fact it gets so hot and dense that the hydrogen starts to ignite with nuclear fusion (↗2). Lighter atoms are fused into heavier ones, releasing high-energy radiation, making it even hotter. Soon, the whole inner core of the gas cloud is burning with a roaring nuclear fire. The outer parts, where the gas is not dense enough for the nuclear fusion to ignite, is not burning, however.

The amount of radiation that gets released whenever e.g. four hydrogen nuclei fuse into a helium nucleus is enormous – it consists not only of high-energy photons, but also of anti-electrons (↗3) (also called positrons) and neutrinos (↗4), which are remnants of two of the four protons transforming into two neutrons. This is what makes nuclear fusion such a great source for energy – and it has two important consequences for the cloud of gas:

Ignition: A Gas Cloud Starts Burning from the Inside

Firstly, it stops contracting. Of course, the pull of its own gravity still exists, but the radiation pressure coming from the regions with active nuclear fusion keeps the gravitational force in balance. There is a delicate equilibrium between the gas collapsing, and the gas constantly exploding! This balance can exist for billions of years, as long as there is enough hydrogen fuel left to keep the fusion process going (↗2).

Secondly, the radiation pressure pushes the outer layers of gas further away. Like a giant leaf blower, the solar wind clears its immediate surrounding of lighter gas particles. This is why the solar system today is relatively free of hydrogen. Even more, with

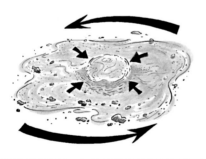

↗2: "Nuclear Fusion" on page 175
↗3: "Antimatter" on page 217
↗4: "The Neutrino" on page 209

modern telescopes one can observe other star systems, and see the regions where their respective stellar winds meet the interstellar matter. Because most star systems actually aren't fixed in the sky, but move through the galaxy, they push a giant tidal wave of hydrogen and helium gas along. This wave, often millions of degrees Kelvin hot, is called the "bow shock", and is a result of a giant nuclear ball of fire ploughing through the interstellar medium.

Planets: Leftovers from Star Creations

Well, so much for the central star, but what about the planets that orbit it? It turns out that the exact details of the formation of our planets is actually quite complicated, and we still have not understood all the details. Roughly it went like this:

Just as the cloud of gas and dirt, out of which the solar system itself formed, was a result of inhomogeneities of densities, so the gas cloud itself is not homogeneous, but certain regions begin to clump together more than others. The biggest one is, of course, its center (the yellow part of the sunny side up), as we have already stated. But also the white part of the egg – called the protoplanetary disc – contains regions which are denser than average, and in these the matter contracts as well, forming clumps, called "planetesimals".

But it seems that gravity is not the only reason for the formation of the planetesimals in the protoplanetary disc. Rather, little dust grains tend to stick together, and form clumps of dirt and rock, sweeping through dust and gas, growing in size as more and more matter sticks to them.

As far as we can tell, the ignition of the central star happened around the time – or slightly before – the inhomogeneities in the protoplanetary disc became prominent, and the matter coalesced into planetesimals. One of the reasons we assume this is due to the fact that – in our star system – the planets closer to the sun, i.e. Mercury, Venus, Earth and Mars, are mostly comprised of heavier elements, which were too heavy to be pushed to the outer rim of the solar system by the solar wind. Also, the planets further away from the sun contain much more gas particles. Jupiter, the biggest gas giant in our solar system, is a good example for this.

After some time however – probably roughly around 4.5 billion years ago – the solar wind had cleared the solar system of most of the gas and dust, so that the growth of the planetesimals came to a halt. That's not when their formation ended, however, because there were still many rocks and debris left over, with which the newly formed planets could interact, either by gravitational force, or by direct impact.

In fact, the dynamics of the planets after their formation is highly complicated, and plays a very important role in the formation of the solar system as it is today.

Genesis of the Moon
A Mini Big Bang Close to Earth

We are not alone in our solar system. There are quite a few neighbors, including other planets that orbit the sun as well. A little closer to the sun we have Venus and a little further away there is Mars, the most popular planet to be visited in the future. The movements of all these planets follow Kepler's laws (↗[1]). But next to all these planets we have a neighbor which is much closer and can easily be seen by eye: the moon.

It orbits our planet at about 1 km/s and takes about 27 days for a full orbit. Sometimes it is right in between us and the sun and causes solar eclipses. It has a radius of about 1700 km, which is roughly one quarter of Earth's radius. But its mass of about $7 \cdot 10^{22}$ kg is only about 1.2% of Earth's mass. Its density is hence much lower. The lower mass causes lower gravitational forces on the moon's surface: instead of with 9.8 m/s^2 (on Earth) you only get accelerated with 1.6 m/s^2, so only about 17% compared to the Earth. This would make you feel pretty light, if you were to take a walk on the moon. Still, the moon is heavy enough to let us feel its gravitational field down here on earth. On the side of Earth facing the moon, the water of our oceans gets attracted. On the opposite side, the water also feels a force away from the Earth's surface. This comes from the relative movement of the Earth and moon around their common center-of-mass. You can try the effect like Emmy and Maxwell pictured here: take a friend's hand and rotate. You feel that your fingers get dragged to the center (by your friend) and your hair, if it's long enough, sweeps to the outside. The same happens to the water on Earth. And if you imagine Earth rotating, every place on Earth will pass by the tidal wave twice a day and we can observe it as ebb and flow tide.

Our Moon – Somewhat Special

Our moon is not only special as it causes tidal waves, bright reflections from the sunlight which can guide you the way home on a dark night and maybe have some other special effects during a full moon. It is

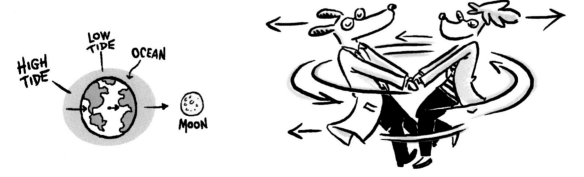

↗[1]: *"Kepler's Laws" on page 47*

also unique in our solar system! Some planets have no moon at all. Others have neighbors that are not really worth being called moons, but are more like asteroids. Our moon is relatively large. It has a lower density than earth, but the elementary composition of its surface rocks is similar to the composition of our Earth. The question is: why is that? And this question is closely related to the question: what is the origin of our moon?

A Collision in Outer Space

Let us go back to the time when our solar system formed. It was, at the very beginning, nothing more than a cloud of dust. Compressed by its own gravity, this dust formed massive objects, starting with the sun and then also with farther large objects emerging from the remaining dust orbiting the sun. You can read more about the birth of our solar system and the formation of planets in a dedicated chapter (↗2).

Before our planet reached its final size, at about 90% of the mass that it has today, it was accompanied by another planet. It had the size of the planet Mars and shared an orbit with our Earth.
That second planet, today also referred to as Theia, had a constant distance to our earth. There are only certain places where such a constant distance can be achieved. These points are called "Lagrange Points".

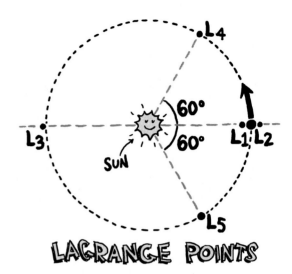

LAGRANGE POINTS

Think of the orbit of Earth around the sun. Let us assume that another, not too heavy object, wants to share that orbit. The problem is that the orbit is determined by the gravitational force between sun and earth. Whenever a third body comes into play, it does not only feel the attraction of the sun, but also the one from Earth. But there are exactly five points in the orbit where the gravitational forces are balanced and keep the orbit of the third body stable. These Lagrange points can be seen in the above illustration. Three of them are on the line between sun and earth.

These Lagrange points are quite popular. Imagine that you want to place a satellite in a spot close to earth in order to observe the sun, such as the SOHO telescope. The best place is at the Lagrange point L_1! And if you are a satellite, and want to stay close to Earth but also want to observe the outer universe without being disturbed by our

↗2: *"Birth of the Solar System" on page 59*

sun: take L_2! This is what the WMAP satellite does as it collects information about the cosmic microwave background ($\nearrow^3$). And if you have a big secret and want to hide it, bring it to L_3. Nobody will ever see it from Earth, because it will always stay hidden by the sun.

Now, let us go back to our Earth prototype and its neighboring planet Theia. It was placed at L_4 or L_5 with constant distance to Earth. But this whole idea of the Lagrange points is only valid for two heavy objects (earth and sun) and a third light one. As Theia grew by collecting more and more dust, its place at the Lagrange point got unstable. At when it had become about as heavy as Mars – roughly one tenth of Earth's mass – it started to move towards the earth.

And then it happened: Earth and Theia collided. Theia was ripped apart and its remnants, together with parts of Earth, surrounded Earth. Theia's iron core, however, merged with Earth. The bits and pieces surrounding Earth formed our moon and our planet gained a little mass. This scenario of the origin of the moon is only one amongst several others. Other ideas are for example that the moon came from somewhere else, accidentally passing by Earth and then got caught by it. But the collision theory does a good job, in particular, it describes the similarity of the Earth's and moon's elementary composition.

Since the creation of the moon, its interplay with Earth has slightly changed. Each year, the distance between Earth and moon, measured via lasers ($\nearrow^4$), increases by about 4 cm. This is caused by the tides that we mentioned in the beginning. During the gravitational interplay between earth and moon the earth's rotation is slowed down, by the friction of water waves, rolling back and forth under the moon's influence. This means that its angular momentum is reduced. As the total angular momentum of the earth/moon system must be conserved ($\nearrow^5$) the moon's angular momentum is increasing. And due to that, the moon's distance to Earth increases as well. From its birth 4.5 billion years ago to now, the distance between the moon and Earth has increased from less than 100,000 km to 380,000 km, which is quite impressive. This means that at the beginning the tidal effects on earth must have been much stronger. Also, as Earth spins slower and slower, our days become longer. But as they increase by only 15 microseconds per year, the effect is nothing that we will realize during our lifetime. Except for one of these days where a leap second is added to our day, as it happened on the 30th of June in 2015.

$\nearrow^3$: *"The Cosmic Microwave Background" on page 105*
$\nearrow^4$: *"Lasers" on page 19*
$\nearrow^5$: *"Conservation Laws" on page 51*

"I'm afraid it's true. Earth is a cosmic franchise."

Extrasolar Planets
Is Anybody out There?

What are we made of? Where do we come from? And where will our journey go? A lot of scientists try with high motivation to answer these questions. While particle physicists try to find out our composition and the way our building blocks interact, astrophysicists want to know if we are alone in the universe or if life exists on other planets as well or is at least possible. Maybe, in the far future we will be able to colonize other planets. Who knows for how long Earth will stay as comfortable as it currently is?

We check our neighboring planets with care, measure their properties and look for signs of life. We search for water, which is already a good indicator, and check the temperatures on the surfaces of our neighboring planets. We found some ice on Mars, for example. The water which was found at the polar caps of Mars is frozen, and that's because of the temperature of Mars. Being further away from the sun makes it colder than Earth: the mean temperature is -67 °F. What about our neighbor on the other side, Venus? Not too cold, but with 867 °F also not very comfortable for life. So it seems that there is not too much life going on on our neighbors in the solar system. Some expect life in the oceans under the ice on Titan, the largest moon of Saturn. But next to that candidate, that is not much promising.

Searching a Second Earth

But why should we restrict ourselves to our own solar system? The estimation of the total number of stars in our universe is about 10^{22}. You can write down the number with all the 22 zeros to see how large it is. Of all these stars, several billion can be seen from Earth. How many depends on your telescope. So why not check these stars for planets? There must be planets orbiting those stars as well! Planets outside our own solar system are called extrasolar planets or exoplanets.

The best way to observe a planet is to take a look at it and see it with our eyes. This is called direct observation. Unfortunately, this turns out to be very difficult. Unlike their star, planets do not emit light directly. The only way they make themselves visible is via the reflection of starlight. While some planets, such as Mars, can be seen with the bare eye, others are very difficult to find, even with telescopes. The planet which is the furthest away from Earth, Neptune, was first observed in a passive way. Distortions of Uranus' orbit, the planet which was the furthest known planet at that time, could only be explained with a yet unknown nearby planet: Neptune. Today we can see him, but only with telescopes. So Neptune, which is "only" about 30 AU (1 AU is an astronomical unit, corresponding to the average distance between the sun and Earth, so about 90 million miles) away from Earth, is already so hard to see. You can imagine that it's also not too easy to see a planet of a neighboring star system. The closest one we know, Alpha Centauri (which is actually a system of two close-by stars), is already 276,110 AU away. Even though that planet would only have to reflect the light of its own star to us, it is still very difficult. So you have to make use of other techniques. We will explain the most popular ones.

Make It Shake

The problem with the extrasolar planets is not only their faint passive light. It is also the relatively bright light of their stars which superimposes onto the planet's own light. In case that the star is a brown dwarf (↗[1]), whose light is not very bright, a direct observation with a good telescope is still possible. But what to do in the case of bright stars? Let us take a closer look at their light. Is there maybe a way that the planet influences its star's light? There is.

STAR MOVING INTO DIRECTION OF EMITTED LIGHT

high frequency

In our heliocentric models we always assume a star is in the center of the system, and planets orbit around it. But this is only true in case the star is by far more massive than its surrounding planets. If there is a very heavy planet In that system, it will constantly pull on the star, so it will not stand still. It also gets attracted by the planet and in the end, both of them surround a common center of mass. For a distant observer this looks as if the star would shake a little, periodic movements both to the left and the right as well as to the front and the back. Now think of observing the moon (which is safe to observe by eye, in contrast to the sun). What could you see more easily, left/right or the front/back movements of the moon? Right, you would see the left/right movement. The

two-dimensional projection that we see makes it hard to see the front/back movements. Now, the situation for a star, shaking because of the movement of a surrounding planet, is different. Here, the resolution of our telescopes is not sufficient to see left/right movements. But instead, these telescopes can use a trick to see both front/back and left/right movements, or more general: changes in its radial velocity. The Doppler shift (↗[2]) leads small variations of the emitted light's frequency while its emitter (the star) moves towards us and away from us. So if we carefully check the frequency of the light, emitted by a star, and there are periodic fluctuations of its frequency, then we know that it is slowly shaking. And this then tells us that there must be another close-by planet. This method, which is called "Doppler method" or "radial velocity method", works best for heavy planets, very close to their stars. An example for a discovery with this method is the observation of the gas giant TrES-4, which has 1.7 times the mass of Jupiter (heavy!) and revolves around its star in only 3.5 days (fast!).

STAR MOVING AWAY FROM EMITTED LIGHT

Low frequency

Mini Eclipses

Another method used for the indirect observation of planets is the "transit method". In the case where an extrasolar planet orbits its star and passes it in such a way that it is between us, observing the star, and the star itself, we get something like a mini eclipse. As we are quite far away from the planet/star system and the star is usually much larger than the planet, the part of the star, which is covered by

↗[1]: *"Spectral Classification" on page 73*
↗[2]: *"The Doppler Shift" on page 15*

the planet, is typically very small. But still, this leads to a small reduction of the intensity of the star's light that reaches our telescopes. If you check how large the reduction is and how long it takes, you can get some information about the planet's velocity and its size. As the variations of the starlight intensity is very small, you need modern telescopes with high intensity resolutions. The Kepler telescope, which started operation in 2009, is able to perform such measurements. You can see the transit method in our illustration.

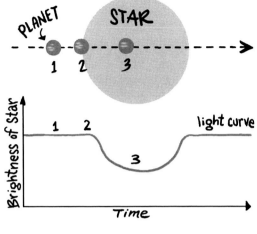

Planets in the Comfort Zone?

About 2000 exoplanets have already been discovered. But you do not simply want to count them – instead, you want to find out what their properties

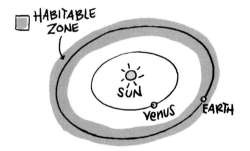

are. Are there maybe some where humans would feel comfortable? This would be a so-called "terrestrial planet", one that has a solid surface and is

not just a ball of gas. It's also important to note the planet's distance to its star. As we mentioned earlier, it can be too hot if it is too close or too cold if it's too far. The zone which is "just right" is called "habitable zone". It is the place where water can exist in its liquid form. In our solar system, only Earth lies within this zone. There are also definitions of the habitable zone which are less strict ("not the best place to live, but some kind of life might still be possible") and cover also the orbit of Mars. Some planets might also have orbits that pass the habitable zone only partially. For an ideal study of an exoplanet's atmosphere and chemical composition, one would have to analyze the planet's emitted light with high precision, in particular its wavelengths. The presence or absence of certain wavelengths is a hint for the emission and absorption of light by

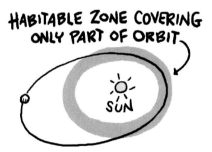

certain atoms. As the technology for the search of exoplanets evolves rapidly, we can expect a lot of interesting discoveries in the future!

II – The Cosmos

"The universe is vast: the sun, which our Earth is orbiting around, is only one of hundreds of millions of stars in our Milky Way. And our Milky Way is only one of hundreds of millions of galaxies in the universe. If you look up at the sky on a clear night, you can catch a glimpse of its marvelous beauty.

And what we can see is only a tiny fraction of what is out there. In the last century, we have gotten better and better telescopes, and have been able to see ever more and more wonders in the universe. The stunning flashes of supernova explosions, glimmering white dwarf stars, compact neutron stars surrounded by vast expanses of brightly shimmering nebulae, all-consuming black holes, and even the afterglow of the big bang itself. All the beauty and mystery of an ever-expanding universe – it is out there, and it will never cease to amaze us. As we will never stop to discover more fascinating and mind-boggling things of the universe we are such a tiny part of.

On the following pages, we will show you a small fraction of the wonders of our universe – the part we know of. Who knows what else is out there, just waiting to be discovered."

*"...and how, exactly, did you harness the heat of the
biggest star in the universe?"*

Spectral Classification
A Who Is Who of Stars

There are many different kinds of stars in the universe. When telescopes became better and better in the beginning of the 20th century, astronomers found that they come in different kinds, depending on how big and bright they are, and what color they have.

Today, astronomers often arrange stars along the so-called Hertzsprung–Russell diagram. This is a diagram where the brightness and the temperature of the stars are shown. The ones to the very left are as hot as 30,000 K, while the ones on the furthermost right are only measly 2,000–3,000 K. In between the stars carry a stellar classification, ranging through O-B-A-F-G-K-M (To remember: "Oh, be a fine girl, kiss me!"). The temperature is directly connected to its color: the hottest O-class stars are a bright white-blue, while the coolest M-class stars are deep red.

The brightness goes from top to bottom in the diagram: if a star is on the very bottom, it is 100,000 times fainter than our sun. If you'd replace the sun with it, daytime on Earth would not be much brighter than under a full moon. On the top of the diagram, on the other hand, are stars which are up to a million times brighter than the sun.

On this diagram, the stars in the Milky Way (and other galaxies, if they are not too far away to identify them), find themselves mostly on a line from the bottom right to the top left. This is not too surprising: for these so-called main sequence stars, the hotter they are, the brighter they are. About a century ago, when this diagram was first set up, it was thought that stars early in their life were very bright, and would become fainter and cooler, as they were thought to shrink down (that was before the 1930s, when physicists knew what nuclear fusion was). This is why the hot, bright ones on the top left are sometimes called "early stars", while the ones on the bottom left are known as "late stars".

But actually, what's going on in the stellar bowels is much more complicated, which is why there are many different kinds of stars. Also, stars change during their lifetime, and one can describe what happens to them by describing how they "wander" through the diagram during the course of their life. This movement takes millions to billions of years, so we cannot see it happening directly. But, since there are so many different types of stars of so many different ages, we can see them at all stages of their existence at the various places of the Hertzsprung–Russell diagram.

Image: ESO

In fact, a normal star usually wanders through the main sequence during its lifetime, and it becomes brighter and hotter, not the other way round. But as a star grows older, it usually does not reach the top left – that is where only the most massive ones are, and they often have already begun their life in that vicinity. And even those ones don't stay there forever! Rather, at some time, almost all stars at some point turn around and wander up and right (or mostly right, when they have already been very far up).

As one can see very prominently when one looks at the Hertzsprung–Russell diagram, about halfway along the main sequence, there seems to sprout an additional "arm", which extends to the top right. Here we find the very bright and relatively cold stars. Actually, this is where the normal and average-sized stars grow old. Here, they grow incredibly in size and become giants, very large and bright, but very red, and therefore relatively cold stellar monstrosities, in the last few million years of their existence.

So an average star will travel along the main sequence from the bottom right to the top left during its life. At one point, however, when it becomes older, it will leave the main sequence and turn to the top right, becoming a giant. At the very top of the diagram, at high luminosities, and with almost all kinds of temperatures, is where we find the giants and hypergiants among the stars. Those are the very massive stars, which have started their lives already very far on the top left, and at some point have started wandering to the right.

What triggers the change of "course" in the diagram? We'll talk about this in detail in the chapter about supernovae (↗[1]).

Crime: Raiding of the Hydrogen lines.

LAST SEEN CANIS MAJOR

Name: Sirius A
Bayer designation: α Canis Majoris A
In constellation: Canis Major ("Big Dog")
Type: A-class giant
Spectral class: A1V
Surface Temperature: 9,940 K
Mass in sun masses: 2
Distance: 8.5 ly
Radius in sun radii: 1.7
Age: 200–300 million years
Fun fact: The brightest star in the night sky, accompanied by a faint dwarf star Sirius B

↗[1]: *"Supernovae" on page 83*

Name: Polaris / North Star

Bayer designation: α Ursae Minoris

In constellation: Ursa minor ("Little bear")

Type: F-class variable star

Spectral class: F7I

Surface Temperature: 6,000 K

Mass in sun masses: 4.5

Distance: 325–435 ly

Radius in sun radii: 46

Age: 70 million years

Fun fact: Actually consists of several distinct stars close together (values shown for the main star).

Name: Menkib

Bayer designation: ξ Perseis

In constellation: Perseus

Type: O-class giant

Spectral class: O7.5III

Surface Temperature: 35,000 K

Mass in sun masses: 30

Distance: 1200 ly

Radius in sun radii: 14

Age: 2–3 million years

Name: Rigel
Bayer designation: β Orionis
In constellation: Orion
Type: B-class supergiant
Spectral class: B9I
Surface Temperature: 12,000 K
Mass in sun masses: 20–24
Distance: 800–960 ly
Radius in sun radii: 70–85
Age: 8–10 million years

Name: Alpha Centauri
Bayer designation: α Centauri
In constellation: Centaurus
Type: G-class dwarf star
Spectral class: G2V
Surface Temperature: 5,790 K
Mass in sun masses: 1.1
Distance: 4.4 ly
Radius in sun radii: 1.2
Age: 5–7 billion years
Fun fact: Consists of two very close by stars (values for α Centauri A – α Centauri B is of nearly the same size, but is a K-class star).

Name: Arcturus
Bayer designation: α Boötis
In constellation: Boötes
Type: K-class orange giant
Spectral class: K0III
Surface Temperature: 4,300 K
Mass in sun masses: ≈ 1
Distance: 36.5 ly
Radius in sun radii: 25.4
Age: 600–860 million years
Fun Fact: The brightest star on the northern hemisphere.

Name: Betelgeuse
Bayer designation: α Orionis
In constellation: Orion
Type: M-class red supergiant
Spectral class: M2I
Surface Temperature: 3,100–3,600 K
Mass in sun masses: 8–20
Distance: 550–800 ly
Radius in sun radii: 900–1200
Age: 7.3 million years
Fun Fact: It pulsates, changing its size, temperature and brightness, with a period of about five years.

Red Giants and Planetary Nebulae
The End of a Main Sequence Star

Stars don't live forever. What happens at the end of a star's life depends very much on how heavy it is. The reason for this is that nuclear fusion (↗[1]) happens quite differently in very heavy and very light stars.

If a star is more than roughly eight times as massive as our sun, it is a hellishly burning furnace, pressure and temperature rising the deeper you are beneath its surface. While hydrogen is fused to helium in the outer shells, the deeper inside you go, the more heavy elements are fused together, up to iron. This is a multi-layered roaring nuclear oven, which, at the end of its lifetime, goes out in a gigantic class II supernova explosion. (↗[2])

But although that is certainly the more spectacular way to go, a majority of stars in our galaxy – about 97% actually – are not that heavy. For the regular stellar Joe in our galaxy, the story is different than for the heavyweights. Make no mistake, stars like our sun are hot – you'd burn up in an instant if you came just near it. But compared to, say, the interior of a B-class supergiant like Rigel (↗[3]), the inside of a typical star in our galaxy is like a refreshing breeze.

Nuclear Fusion in Lighter Stars

Firstly, the pressure and temperature in stars like our sun is not high enough to actually have nuclear fusion happening everywhere. Rather, the actual fusion of hydrogen into helium happens only in the very core, in a region between the center and about a fifth of the way to the surface. Here the star consists of a mixture of hydrogen and helium, and the actual burning is happening here, at a comfy temperature of a few million Kelvin. Everything further away from the center is just hydrogen plasma – certainly hot enough to fry you in an instant, but not hot enough to undergo fusion.

Stars of Old Age: Red Giants

Now what happens when such a star grows old? The first change certainly occurs when all of the hydrogen in the core region is burned up, and only helium is left. The star, at this point, will most likely not be massive enough to start the next fusion step – the burning of helium to heavier elements such as carbon and oxygen. So the core will just stop burning, and begin contracting under its own weight. There is still more hydrogen left in the regions not as deep down of the star, so the nuclear fusion will continue – in a ring of fire around the core, eating its way outwards, burning up the rest of the star.

It is at this point when a star typically leaves the so-called main sequence and begins wandering upwards in the Hertzsprung–Russell diagram (↗[3]). The burning layer of hydrogen which slowly wanders

---

↗[1]: *"Nuclear Fusion" on page 175*
↗[2]: *"Supernovae" on page 83*
↗[3]: *"Spectral Classification" on page 73*

to the surface expands the star dramatically – this makes the star much much brighter and larger, but also slightly cooler. It becomes a red giant. When this will happen to our sun (in a few billion years), its radius will increase to about 200-fold, swallowing the inner planets Mercury, Venus and Earth.

Helium Burning Finally Sets In

At some point during the red giant phase, it will happen that the core has contracted enough, and temperature and pressure have risen enough so that the next fusion step – from helium to carbon and oxygen – ignites. This needs about 300 million Kelvin in the core, while in the outer regions the temperature will just be a few thousand Kelvin.

For slightly heavier stars this happens earlier, for the very light ones it happens much later, when a lot of helium from the outer burning shell has been gathered. In the former case, the burning starts slowly and in a controlled fashion, in the latter case it happens abruptly in the so-called helium flash.

Either way, the expansion of the star is momentarily stopped as the helium burning sets in, but as soon as all of the helium in the very core is used up and

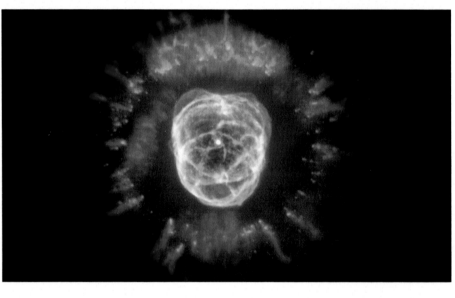

a non-fusing carbon/oxygen core remains, together with a shell of burning helium slowly wandering outwards, the growth of the red giant commences.

The Hiccup Phase: Stellar Winds Rip the Star Apart

At that time, the burning hydrogen layer will not have reached the surface of the star. Now, with two expanding layers of nuclear fusion, it will become more and more unstable. You see, a star is not a very regular thing, where everything happens decently and neatly confined to its own layer. Rather, it can be a quite turbulent matter, where the plasma flows back and forth inside the star. If it just consists of a burning core and the rest, then gravity keeps everything together most of the time. But, the more intricate things go on in its interior, the more unstable it becomes.

Image (left):   NASA, ESA, Andrew Fruchter (STScI), and the ERO team (STScI + ST-ECF)

Image (right):   NASA, J. P. Harrington (U. Maryland) and K. J. Borkowski (NCSU)

This is why, at this time, the star goes into the hiccup phase. While this is not official terminology, it probably describes very well what happens: Regions of stellar fusion ignite and go out at various regions in the star, and the whole burning process becomes very turbulent and irregular. The actual details can only be simulated with high-end computers nowadays, but the result is clear: Massive stellar winds rip the outer regions of the star apart, and swathes of stellar matter – hydrogen and helium alike – are

the core itself often continues to emit radiation of some sort (↗⁴), this lights up the gas clouds (although mostly not in colors the naked eye can see) surrounding it. These clouds are called planetary nebulae, although the name is historical, and they have nothing to do with planets. You can see two of such beautiful examples in our images. The first one shows NGC 2392, the so-called "Eskimo Nebula" in the constellation of Gemini. The second, on the right, is NGC 6543 in the constellation of Draco. It is called "Cat's Eye Nebula", and this is a famous image combining X-ray and visual light.

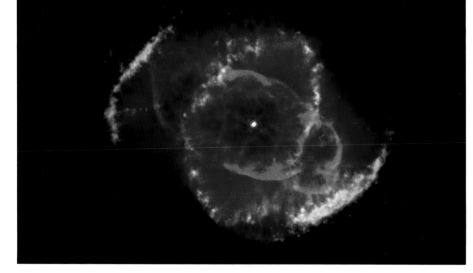

The creation of a planetary nebula at the end of a star's life plays an important role in the whole evolution of the galaxy. It returns vast amounts of helium and hydrogen to the interstellar medium, the gas and dust between the stars. This material can and will later be attracted by

jettisoned into space. The hiccup has become an outright vomit.

other, more dense regions, and eventually form new stars (↗⁵).

A Planetary Nebula Is born

Although its sounds nasty, the result is actually one of the most beautiful phenomenon in the galaxy: the ejected matter surrounds the remaining core of the star like a giant cloud of ethereal matter. Since

And what remains? Well, the former core of carbon and oxygen is all that's left of the star. It will be very compact, but it will not undergo nuclear fusion. It will still be quite hot, and slowly cool off over the next million years. This is what's called a white dwarf, and it will get its own article in this book (↗⁶).

↗⁴: *"Alpha, Beta and Gamma Rays" on page 171*
↗⁵: *"Birth of the Solar System" on page 59*
↗⁶: *"White Dwarfs and Type Ia Supernovae" on page 87*

"Do they realize it's six hundred and forty light years away?"

Supernovae
Going out with a Bang

Nothing lasts forever, not even the stars in the sky. Actually, stars are born (↗[1]) and die all the time in the universe. And when they end their life, they don't go quietly. They light up in one of the most violent and energetic phenomenon in the universe, when for the duration of weeks, it becomes brighter than entire galaxies, before fading away. This is called a supernova explosion.

There are different types of supernovae, called type Ia, type Ib and Ic, as well as type II (Footnote: there are other types as well, but they do not occur very often). They differ in what kind of radiation is emitted during the explosion. Which type of supernova happens when the star's time is over depends very much on how heavy it is.

Starfire: Nuclear Fusion on a Stellar Level

When a star is burning, the hydrogen in it is converted to helium – this process, of fusing lighter elements to heavier ones, is what keeps the star burning. The radiation pressure of nuclear fusion (↗[2]) keeps the star stable against its own gravity. In fact, one could rightfully say that stars are constantly exploding and collapsing at the same time! But the fuel inside a star does not last forever, and when it runs out, a star undergoes changes, sometimes very rapidly.

Most stars run out of hydrogen in its center first, where the temperature and pressure is high-est. That might not be the end of the story yet, though, because under great pressure the helium can fuse into even heavier elements, such as carbon or neon. These processes do not give as much energy as the hydrogen fusion, but they give at least some. These heavier elements can fuse into even heavier ones, and so forth, up to iron. Iron nuclei are the energetically most stable ones. That is why you can get energy by fusing together nuclei lighter than iron, and splitting apart those heavier than iron (↗[2,3]).

The Iron Heart of a Star

Anyway, because of this cascade, many stars actually are a bit like many-layered onions: on the outside shell hydrogen is still burning, below that there is a layer where helium is burning, and so forth, until you reach the center of the star. A young or rather light star might only have few layers, while an old, or very massive star, can actually have many layers, and have quite heavy elements burn up further inside. Heavy stars will have extremely dense cores of iron in its centers.

Such an iron core cannot undergo fusion anymore, so it has to hold up the weight of the entire star pressing down on it by itself. The iron atoms are packed together tightly, compressed into an extremely dense region by its own weight, as well as by the weight of the outer, still burning layers of the star. Most of the resistance comes, in fact, from the electrons in the iron: electrons are fermions, and that means two of

↗[1]: *"Birth of the Solar System" on page 59*
↗[2]: *"Nuclear Fusion" on page 175*
↗[3]: *"Radioactive Decay" on page 167*

them cannot be in the same state – roughly speaking, where one electron is, another cannot be ($\nearrow^4$).

The End Is near When the Iron Core Grows Too Heavy

As more and more fuel in the next layer runs out, and more iron is produced, the core slowly gains in mass (shrinking slightly due to the increasing gravitational force, because it gets even heavier). At some point, however, the pressure becomes too much: the cores become so massive that the electrons get pressed into the protons of the iron nuclei. In a core-wide β^+ decay with electron capture ($\nearrow^5$), the innermost regions of the star undergo a catastrophic change. Electrons and protons turn into neutrons, which release highly energetic radiation during this process. Like a landslide, iron cores are reverted to simply neutrons, the energetic gamma rays being set free by this process shred through the core, fracturing its nuclei even further. Now, neutrons can be packed much tighter than electrons and iron nuclei. The immense weight and pressure contract them all into a tightly packed ball – within seconds, the whole stellar core becomes a neutron star ($\nearrow^6$). The outer layers of the star have not yet realized what has happened: the iron core of the star has nearly vanished into a tiny lump of neutron matter. But inevitably gravity also takes hold of them: as there is nothing to hold it up, the whole star, from the innermost to the outer layers, collapses.

The Collapse: From Iron Core to Neutron Star in Seconds

Meanwhile, the core has coalesced into a nearly perfectly round ball of tightly packed neutrons, again hold up by the fact that neutrons are fermions: Wherever there is one, another cannot be. Although the neutron core is tiny compared to the rest of the star, it is basically a gigantic atomic nucleus, just without any protons. That also means it is incredibly hard – and now the rest of the star crashes onto it!

With supersonic speed, the outer layers of the star impact onto the rebounding neutron core. This generates a massive shock wave through the stellar matter, and it is this shock wave which carries what is left of the star outwards. The shock wave is supported by the large amounts of neutrinos ($\nearrow^7$), which have been generated in the beta-plus decay. Usually, neutrinos pass through matter nearly unhindered. But in the catastrophic core collapse, a lot of them have been generated. And they have a lot of energy. With tremendous force, the neutrinos wash the star remnants away from the core – the star explodes one final time, shooting stellar matter and neutrinos into the universe.

Why do we know this? Well, although there are still some unclear parts about the details of the mechanism, the collapse of a star is understood so well because we have incredibly precise computer simulations of this. For instance, it is because of these simulations that we think the neutrinos themselves play such an important role in generating the massive supernova explosion. For quite a while, physicists assumed that the impact shock wave would just travel through the neutron core, coming out the other side and carrying the matter away. It is just after the detailed computer simulations that they realized the shock wave alone is not enough to generate these enormous supernova explosions – but that the neutrinos help with it.

$\nearrow^4$: "Spin" on page 187
$\nearrow^5$: "Alpha, Beta and Gamma Rays" on page 171
$\nearrow^6$: "White Dwarfs and Type Ia Supernovae" on page 87

$\nearrow^7$: "The Neutrino" on page 209

Now all we have said so far applies to rather massive stars – roughly above 8 solar masses. All supernova types except type Ia happen because of a core collapse. The process we have described above is pretty much what happens for a type II supernova. But also type Ib and Ic supernovae have collapsing cores – it's just that they have, in their lifetime, already lost much of their outer layers, possibly by strong solar winds, or because a companion star has stripped it away with its own gravity. Because much of the hydrogen and helium layers are missing, the radiation looks characteristically different, which is why one can distinguish them from type II supernovae.

Type Ia supernovae however, they are completely different beasts. They are exploding white dwarfs, and will be tackled in the respective article (↗6).

in our Milky Way, on average about three supernovae occur per century. There have been very violent ones in human history, which have been recorded. The earliest recorded observation of a supernova was is 185 AD by Chinese astronomers. Nowadays it is called SN185, and this naming convention has been kept for all records. One of the brightest ones must have been SN1054, which was visible with the naked eye during the day for about a month, and during nighttime for nearly two more years after that. The remnants of this explosion form the Crab Nebula, near the constellation of Taurus. The most recent one to be seen directly was SN1604, which was studied in detail by Johannes Kepler.

The supernovae are not just destructive, though: the stellar matter ejected contains many higher elements, from carbon, over oxygen, up to iron, which are essential for the formation of planets that sustain life (↗1). It is very likely that all atoms in your body have originally been created in a star, and carried off into the universe by a supernova explosion.

Supernovae in History

Supernova explosions do not happen all too often. From observations in other galaxies, we suspect that,

Image: NASA, ESA, J. Hester and A. Loll (Arizona State University)

White Dwarfs and Type Ia Supernovae
Corpses of the Suns and Standard Candles

In other articles we have talked about the final days of a star's life. But after the violent expulsion of stellar matter into the universe (↗1) or the even more dramatic catastrophes that are the supernovae explosions (↗2), what is left? What remains after the gamma radiation bursts diminish, and the last clouds of hydrogen are blown off into the void?

The Fate of Typical Stars: the White Dwarf

We have already hinted at it in the respective articles: whenever a rather light star, such as our sun, ends its life, the remainder is what's called a white dwarf. The outer shell of the star, the hydrogen and helium, have been ejected into space, and the core of the star is exposed. This core consists of carbon and oxygen atoms, which are hold together tightly under their own weight.

White dwarfs often have a mass between half, and a bit above, the mass of our sun. However, they usually are not much larger than the Earth. In other words, they are enormously dense! A teaspoon

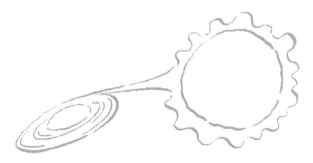

full of white star matter weighs about as much as a car (and not one of the fuel-efficient, light ones). Since it is so heavy, the gravity force on its surface is enormous: any person standing on a white dwarf would be immediately squished to a fine, thin layer around it.

The Warm Nuclear Afterglow of the Stars

While that happens, such a person would also immediately burst into flames, because a white dwarf is hot. The fusion processes have stopped, but the white dwarf matter has still stored a lot of heat energy from its time as the core of a star. This heat is given away only very slowly; often, a white dwarf has a temperature of many tens of thousands of Kelvin, and cools off over the course of million years.

The reason why a white dwarf is not collapsing any further, is that the electrons and atomic nuclei of its carbon and oxygen atoms are tightly packed. In particular the electrons play an important role here: they are so-called fermions. By the Pauli exclusion principle, two fermions cannot be in precisely the same quantum state. In a way, they are like solid little balls: where one is, another cannot be. Footnote: this is not precisely true: two electrons are allowed to be at the same place, if they have different velocities, or different spins. But at some point all types of electrons with all types of velocities of a certain energy are present, so one cannot add another to the same place. This is what keeps them apart. This is different from bosons, like photons: there is no

↗1: *"Red Giants and Planetary Nebulae" on page 79*
↗2: *"Supernovae" on page 83*

problem of piling up as many of them to a point as you like.

So what's keeping the white dwarf stable is quantum mechanics. More specifically, it's the Pauli principle. And it is doing a great job at it: the galaxy is full of white dwarfs – in the Hertzsprung–Russell diagram (↗³) they have their own region, far below the main sequence. They are usually hot, but very faint compared to a normal star. But one can see them glowing. The white dwarf nearest to us is Sirius B, the companion of the A-class giant Sirius A. It is only about 8.5 light years away from us, about as heavy as the sun, slightly smaller than the Earth, and has been cooling off to a cozy 25,000 Kelvin for the last 120 million years.

Growing White Dwarfs: Type Ia Supernovae

Now, white dwarfs are usually not all alone in the galaxy. No star really is: there is lots and lots of gas, dirt, interstellar matter, which swirls around in clouds. Sometimes it hits a white dwarf. Sometimes a white dwarf is the companion of another, still burning star, which constantly showers his small partner with stellar matter, coming from solar winds,

eruptions, or else. Any way, there are a million ways to acquire more mass in the galaxy. And for a white dwarf, at some point, there is a point where it cannot hold off any more: with enough mass it can build up enough pressure to ignite fusion again. What usually happens in the inside of very massive stars – the fusion of carbon and/or oxygen to iron – now happens instantly. A soon as the white dwarf has acquired enough matter, it ignites. All at once. Like a gigantic nuclear bomb, all of its matter undergoes fusion instantly, and the star rips apart.

For a moment, the white dwarf shines brighter than a whole galaxy. This is a very specific type of supernova explosion, which has been called "type Ia", for historic reasons (↗²). The interesting point is that these explosions more or less all look the same – while normal supernovae come in various types and sizes, supernovae of type Ia are very characteristic, because it is always a very specific amount of carbon and oxygen exploding.

Standard Candles: a Way to Measure the Universe

This is unfortunate for the white dwarf – but fantastic for astronomers here on Earth! It means that type

↗³: "Spectral Classification" on page 73

Ia supernovae can be used as so-called standard candles. You see, it is not easy at all to measure

how far something is away from us in the universe. In particular with galaxies further away, one cannot really be certain of whether it is so faint because it is far away, or because it is so small, or because there is so much dust and dirt between us and them, obscuring it for our telescopes. But because type Ia supernovae are so very typical in their brightness and spectrum, they can easily be recognized, and used to measure the distance to the galaxy where it originated from.

In the past decades, this has been used to measure very precisely where the other galaxies are in relation to us, and in particular how far away they are. These measurements have been paramount in realizing that the universe is, in fact, expanding in an accelerated fashion, driven by the mysterious force of dark energy (↗4).

The next Stadium: A Neutron Star

What remains of a white dwarf is usually not simply an iron core, although that can happen in some cases. More often, though, the white dwarf has

acquired more than 1.4 times our sun's mass, and this is beyond the so-called Chandrasekhar limit: It is the maximum weight of an iron core. Heavier than that, and it cannot keep itself up like the white dwarf did. Rather, the electrons are pressed into the protons, which leaves only neutrons. The remainder is therefore a neutron star, a giant atomic nucleus. These also occur as the corpses of massive stars, remnants of collapse-supernovae (↗2).

A neutron star is even more extreme than a white dwarf: with a weight of 2–3 suns, and a radius of a few kilometers, it is really hellish in its direct vicinity. It is again kept up by the Pauli principle (yes, neu-

trons are also fermions). It cannot get any heavier than that – if it gets beyond this so-called Tolman–Volkoff–Oppenheimer limit, the sun curves the space around it so much (↗5), that it begins to fall behind its own event horizon. The result is a black hole (↗6), the final stadium of any stellar evolution. Well, at least as far as we know (↗7).

↗4: "Dark Energy" on page 137
↗5: "Curved Space Time" on page 125
↗6: "Black Holes" on page 91

↗7: "Black Hole Evaporation" on page 283

Erwin's Journey through a BLACK HOLE

AS OBSERVERS, WE SEE POOR ERWIN SPIRALING TOWARD THE EVENT HORIZON —THE POINT AT WHICH THE GRAVITATIONAL PULL IS SO GREAT THAT MATTER CANNOT ESCAPE.

TO US, HE APPEARS TO MOVE SLOWER AND SLOWER, BUT IN FACT ERWIN IS ACCELERATING.

ERWIN FEELS THE PULL OF GRAVITY STRONGER AT HIS PAWS THAN AT HIS HEAD BECAUSE HIS FEET ARE CLOSER TO THE BLACK HOLE.

THE BLACK HOLE'S GRAVITY IS **SO** INTENSE THAT LIGHT FROM DISTANT STARS BENDS AROUND IT.

THE CLOSER ERWIN GETS TO THE BLACK HOLE, THE STRONGER THE PULL OF GRAVITY... AND THE FORCE STRETCHES HIS BODY...

DEPENDING ON THE SIZE OF THE BLACK HOLE, ERWIN MAY GET A PEEK BEHIND THE EVENT HORIZON BEFORE HE IS SHREDDED TO SMITHEREENS.

Black Holes
Once You Go Black, You Never Come Back

There are many wondrous and fascinating objects out there in space: gaseous nebulae stretching over dozens of light years, neutron stars, with diameters of a few miles, but weighing many times as much as our sun, or giant red stars, nearly as large as our whole solar system. But no stellar phenomenon in the universe is as mysterious and enigmatic as the black hole.

The theoretical possibility of the existence of black holes as a consequence of Einstein's Theory of general relativity (↗1) had been known for nearly a hundred years. But it wasn't until the 1960s, when physicists came to realize that they are not just freak solutions to the equations, but could actually be present in our galaxy – and might in fact be quite numerous, as they are the burnt-out leftovers of heavy stars.

In the End, Gravity Always Wins

The existence of a star is a constant struggle: while the fire of nuclear fusion inside it threatens to tear it apart, the weight of its own mass pulls it together, aiming to contract it entirely. These two opposing forces – radiation pressure and gravity – usually are in balance with each other. So for billions of years, a star can be a stable source of light and radiation in the universe.

However, the nuclear fuel, hydrogen, doesn't last forever. At some point, when most of it has been converted to helium and heavier elements, the radiation pressure which keeps the star from collapsing,

weakens. Gravity, however, never weakens. The ignition of the heavier elements in the core cannot sustain the massive weight of the star, which eventually collapses. In the end, gravity always wins.

During this contraction, some of the outer layers of the star get ejected into space in a violent explosion, called a supernova. What is left of the star keeps contracting. If the remnant is heavy enough – having a mass beyond the so-called Tolman–Oppenheimer–Volkoff limit of around 1.5–3 times the mass of our sun – there is nothing that can stop it from collapsing completely under its own weight. What remains is commonly known as a black hole.

In a black hole, all of its mass is concentrated to a microscopic point, called the singularity. There is a region around this singularity, where the gravitational force is so incredibly strong that nothing can escape it. No matter how powerful the engines of your spaceship, if it is close enough to the singularity, it inevitably falls into it. The boundary of this region – the point of no return – is called the event horizon of the black hole, and whatever crosses it, is lost

↗1: *"The Theory of General Relativity" on page 121*

forever. Not even light itself can escape the gravitational pull once it passes the event horizon. This is why this phenomenon is called a black hole: there is no way we can see it directly, because it does not emit anything, not even radiation.

We Cannot See It Because It Does Not Emit Anything

So what does a black hole look like? If we were to encounter one, and try to observe it (hopefully from a safe distance), we would simply see a completely black void, a sphere of nothingness where the event horizon is. Moreover, the gravity of the black hole is so strong that it even alters the course of light rays that pass it at some distance: We would see this as a strong distortion of the image of the stars which are, from our point of view, behind the black hole. It acts as a so-called gravitational lens (↗2).

Everything we have said so far is true for a "pure" black hole, a black hole which is completely on its own, without any matter in its immediate vicinity.

In reality, though, a black hole would not be just alone by itself. It would be surrounded by large amounts of matter, caught during its existence: the remnant of its former life as a star, corpses of other stars, the occasional stray asteroid, clouds of gas, and interstellar matter. All of it trapped in its vicinity, crunched and crushed to a spinning, swirling pancake called the "accretion disc", which envelops the black hole.

A piece of matter in the outer rim of the accretion disc can orbit the black hole many times, spiraling further and further inwards, giving off intense electromagnetic radiation in all frequency bands, before it finally crosses the event horizon, utterly destroyed when it finally falls into the singularity.

Of course, an observer from far away would never even see it crossing the horizon (see the extra text "What happens when you fall into a black hole?"). The only thing to see would be the radiating matter in the accretion disc spiraling into the black hole, becoming slower and slower as it approaches the

↗2: *"Gravitational Lensing" on page 129*

event horizon. Also, it would glow more and more in deep infra-red colors, not visible any more to the naked eye.

So, how many black holes are there in the universe? The answer is: nobody really knows. But there are very strong indications that Sagittarius A*, an extremely strong and compact source of radiation in the Sagittarius constellation, in the center of our galaxy, is in fact a black hole, together with an incredibly bright accretion disc surrounding it. The reason why we suspect this, is that we have observed many radiating stars in its direct vicinity. From their extremely fast, circular movements around a common center, one can deduce that there has to be a black hole, of roughly four million times the mass of our sun.

In fact, supported by astronomical observations, astrophysicists suppose that this is the case for most known galaxies: in their center they contain a super massive black hole, around which the whole galaxy revolves.

Apart from that, unfortunately, it is very difficult to say how many black holes there are, even within the confines of our galaxy. The Milky Way contains several hundred billion stars, many of which are massive enough to collapse to a black hole when they run out of fuel. So black holes should be a common occurrence. But there is not a single object right now, of which we can be absolutely sure that it is a black hole, although there are many good candidates. The problem is that, from a distance, it is actually not so easy to spot the difference between a very bright star, and the accretion disc of a black hole. The closest black hole candidates are still several thousand light years away from us.

One of them is Cygnus X-1, a source of immense X-radiation, about 6.700 light years away. There is a famous bet between Kip Thorne and Stephen Hawking about whether Cygnus X-1 is in fact a black hole or not. By now, evidence for a black hole is so strong, that Hawking has willingly conceded the bet, which won Kip Thorne a year's subscription of Penthouse.

What Happens When You Fall into a Black Hole?

Simply put: you die. Sad, but true. But you die in a very interesting way, that is worth talking about a little.

When you are far away from the black hole, falling towards is like free falling everywhere in the universe: you feel weightless. So you do not notice anything, apart from this little black dot coming closer and closer.

When you approach the event horizon, you are still free falling, but you notice something strange when you look at the other stars around you. Their image gets more and more distorted, because light that passes close to the black hole bends around it. The black sphere that marks the event horizon takes up more and more of the sky, seemingly growing large and surrounding you more and more.

Shortly before you cross the event horizon, the last stars can only be seen as if through a narrow tunnel, when you look behind yourself. You are not yet in the black hole, mind you – it's just that most of the light coming from the stars does not reach your eye. Rather, it is sucked into the black hole.

The last bit of normal universe vanishes from your sight, as soon as you cross the event horizon. From then, it only takes moments until you crush into the singularity.

You die slightly before that, though. Although you are freely falling the whole time, at some point the tidal forces become too enormous to survive. A part of your body which is closer to the black hole experiences a slightly stronger gravitational force than a part which is further away from it. At some distance from the black hole, this difference might not be much. However, at some point during your journey to the singularity, this difference becomes so large that you are stretched long and thin, before you get simply ripped apart. This process has been dubbed "spaghettification" by some.

When this occurs exactly on your way to the singularity, however, depends on the size of the black hole. For a smaller black hole, the event horizon is so close to its center, that you'd be killed long before you cross the point of no return. The more massive the black hole, however, the further out the event horizon is. For one as heavy as the super massive black hole in Sgr A*, for instance, one is being ripped apart after one crosses the event horizon. At the horizon itself you would actually still feel quite comfortable.

Interestingly, all of this is only happening from your point of view. A distant observer with a good telescope would make a different observation: She could still see how you approach the event horizon. But as you approach it, she would actually see you become slower and slower, coming closer and closer to the horizon, but never reaching it. She would still see a frozen image of you shortly before being sucked into the black hole, long after you have perished inside it.

The reason for this lies in the nature of the event horizon: Light from behind the event horizon can never escape the black hole, and never reach the eye of our observer. Light from very close in front of it can escape, but it might take a very long time. The light rays sent out from you the second before you cross the horizon, have to slowly crawl up the gravity well for centuries, before escaping the gravitational grasp and traveling to the observer's eye. During this struggle, the light loses a lot of its energy, shifting its color spectrum to the infra-red. So for all eternity, a more and more red-shifted, squashed, still image of you just crossing the horizon would be seen hanging in the sky.

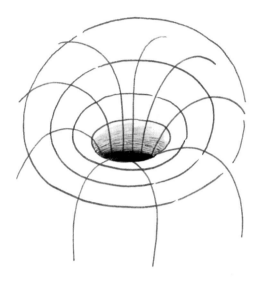

A COOL TRICK

TO SEE THE BLACK HOLE, HOLD THE BOOK AT CHEST-LEVEL A FEW INCHES IN FRONT OF YOU, AND VIEW IT THROUGH YOUR CELL PHONE CAMERA.

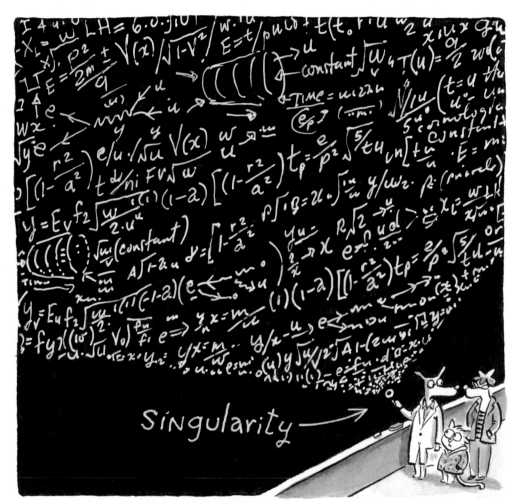

*"We've reached the point where equations break down, space
and time are born, and physicists go to lunch."*

The Big Bang
The Horrendous Space Kablooie

If one looks up in the sky, and tries to fathom the vastness of space out there, beyond the Earth, the solar system, and even our galaxy, one cannot help but feel a sense of wonder and awe. Many people have had this sentiment over the centuries, no matter whether they had telescopes at their disposal, or just a clear look at the night sky. Such a wonderful and marvelous thing as the universe, many thought, certainly has to have a spark of divinity in it. And as is customary for the god in any Abrahamic religion, it should be eternal and unchanging in its magnificence.

That was the dominant view throughout much of history, in most of western culture. It came as much of a shock to Albert Einstein, therefore, when he found out that his recently published equations, which describe space and time ($\nearrow^1$), did in fact not allow for a static and unchanging universe. His theory predicted that the universe either had to be expanding or contracting, but could not stay the way it was. Horrified, Einstein quickly modified his equations by adding an additional term, which he dubbed "cosmological constant". For his modified equations, there were solutions which allowed a static universe, so he was satisfied at first.

Hubble Realizes:
the Universe Expands!

It was in 1929 however, when an astronomer called Edwin Hubble published his findings about the spectral emissions of stars and nebulae. His findings were phenomenal: he had looked at the light coming from different sources in outer space, and had discovered that most of it was red-shifted. That means, that the Doppler effect ($\nearrow^2$) had stretched the wave length of the light sent out by these other stars, during their way to the Earth. This meant that all of these stars had to move away from us! Even better: They were moving faster away from Earth, the further away they were. There was only one explanation: the whole universe had to be expanding!

Wait a second – you might say – why does that mean that the universe is expanding? Maybe all the stars and galaxies are actually just flying away from us. What does this have to do with the universe expanding?

Well, there are two good reasons why Hubble's findings don't mean that the universe is static, but everything flies away from us: Firstly, it would mean that the Earth is exactly in the center of the universe, and that would make our part of the universe very, very exceptional. With such a large universe, which has stars and galaxies everywhere, one would be hard-pressed to find a good explanation as to why everything is trying to get away from our position.

Secondly, the speed in which things are moving away from us increases linearly, the further away objects are. This linearity – which nowadays is

$\nearrow^1$: *"The Theory of General Relativity" on page 121*
$\nearrow^2$: *"The Doppler Shift" on page 15*

called "Hubble's law" – means that far away galaxies can recede much faster than the speed of light! If a galaxy is roughly 4.3 Gigaparsec, or 14 billion light years away, it travels away from us with the speed of light. If it is twice that distance from us, it travels away from us with twice the speed of light, and so on.

The Universe Expands – Even Faster than Light!

If these were just galaxies moving through space, that would clearly be forbidden – nothing can move faster than the speed of light (↗3). But in fact, these galaxies are not moving – rather the space between them is expanding! A good way to visualize this is the inflation of a balloon: The balloon is the universe, in this case. Also, assume that there were lots of little ants crawling around on the surface of the balloon while it expands. To each of them, it would appear as if the other ants were moving away, the faster and the further away they are. The situation with our universe is similar – it does not matter that the ants have a maximum speed with which to crawl over the balloon (that is the speed of light in our analogy), there is no limit to the "speed" with which space is expanding! So again, galaxies only appear to move away from us – in reality, the space between them and us gets bigger. And the rate of that expansion does not have an upper limit (which is why something like warp propulsion can theoretically work, ↗4). And this is

why it appears that far away galaxies are moving away from us faster than light.

By the way: the expanding universe suddenly fit perfectly with Einstein's original equations, so he hastily dropped his cosmological constant, and later called it "his biggest blunder". Nowadays, one uses the equations with a cosmological constant again, but with a different value than the one Einstein used back then.

A long Time Ago: the Big Bang Singularity

But there is more: the theory of general relativity does not only describe the universe as it is today. Also, it tells us that the universe has been expanding for its whole lifetime, and that roughly 13.8 billion years ago, it was very dense and very hot. The equations then cannot be calculated further backwards than this point – the temperature, density and curvature become infinite. That instance is nowadays termed the Big Bang, and many physicists assume that the universe actually began its existence at that time.

That our equations break down at this point is very unsatisfactory, however. In fact, the universe must have been so very hot and dense at that time that many physicists expect Einstein's equations to not be valid any more in that situation. Rather, all the particles in the universe probably were so very

↗3: "Relative Space and Time" on page 117
↗4: "Warp Drive" on page 295

densely packed together that quantum effects of elementary particles (and probably of space-time itself, ↗⁵) would need to be taken into account. There is, at the moment, no satisfactory description of quantum and relativistic effects at the same time, so we do not know what precisely happened at the Big Bang. There are a few things we can be relatively certain of at the moment, though.

GOODBYE!

1.) The Big Bang description of the cosmos is a very good model up until a few tiny fractions of seconds after the singularity. We have a quite good understanding of which particles were created when, and in what amounts.

2.) The Big Bang did not happen somewhere, but everywhere at once. The image that there was an explosion somewhere, and then, from a tiny point, all the matter started filling empty space, is wrong – rather, the universe was very concentrated and hot everywhere.

Big Crunch/Big Bang

3.) One often hears that the universe was "very small" at the beginning. In fact, we do not know this, because we do not know whether the universe is of finite size or infinitely large. If the latter is case, then the universe was always infinitely large, even at the Big Bang. In that case, it was just very dense and hot, and has expanded and cooled down since then. It has

time

grown a lot since then – but has still always been infinitely large.

4.) We do not know what happened before the Big Bang, or even if anything happened at all. There are several possible scenarios, none of which is in any way "certain", or even "likely": Both space and time really have begun their existence at the Big Bang. In that case, the Big Bang was the first thing that ever happened, and asking what came before is like asking what is further north from the North pole – that question does not make any sense.

SEE YOU AT THE BIG CRUNCH!

Another possibility could be that there was something before – for instance some other, collapsing universe. That other universe would have been contracting under its own weight, until it crumpled together in a Big Crunch – from that very dense and hot situation it "bounced back" and started expanding again – to form our present day expanding universe. Even though that would put something before the Big Bang, it might still not be sensible to ask how long that moment took – simply because during the moment of strong contraction there was a large quantum uncertainty to the time duration between collapsing and expanding phase (↗⁵).

↗⁵: *"Quantum Gravity" on page 279*

Because the universe is expanding rapidly, we can be pretty certain that at some point the universe was very small, or rather, that everything was very close to everything else. But in fact, there is more. We have a pretty good understanding how particle creation works, and Einstein's equations of General relativity describe in detail, how various forms of matter interact with curved space-time. Comparing this with the observation from astrophysics in an age of better and more precise telescopes, we can actually get a quite precise idea what happened in the universe, even fractions of seconds after the Big Bang – what happened precisely at the Big Bang still remains a mystery, so far. Let's go back to right after the universe was created, and have a look at what was going on – to the best of our knowledge today.

Let's check the most important properties of our Universe during it's evolution:

Age of the universe. In other words: how much time has passed since the Big Bang?

How hot is the universe at that time, roughly? We give the temperature in K (Kelvin), but you can easily just think Celsius or Fahrenheit – with these high numbers, it actually doesn't matter.

The factor, by which the universe at that time was smaller than the universe today.

What is going on at this time in the universe? What kind of matter, radiation, energy, is around?

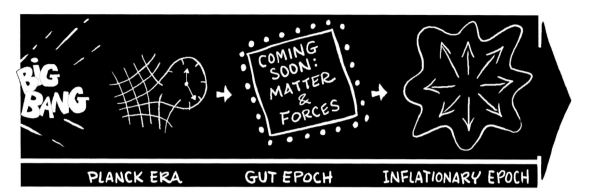

PLANCK ERA	GUT EPOCH	INFLATIONARY EPOCH

0–10^{-43} seconds.
Actually, it is unclear if time is passing at all in the usual sense.

Probably around 10^{32} K

Similarly as time, the concept of space itself probably did not make much sense.

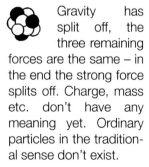 Really not much about the physics at this kind of density and temperature is known. It is suspected that gravity and the three other fundamental interactions are unified, which means they are really just four different aspects of the same force.

10^{-43}–10^{-36} seconds

Probably around 10^{29} K

10^{-51}

Gravity has split off, the three remaining forces are the same – in the end the strong force splits off. Charge, mass etc. don't have any meaning yet. Ordinary particles in the traditional sense don't exist.

10^{-36}–10^{-32} seconds

During this time, temperature drops by a factor of about 100,000, to $\approx 10^{24}$ K.

Here the universe grows incredibly fast, by a factor of at least 10^{26}.

The inflationary era is a phase transition (very much like ice melting or water boiling), and starts as a result of the strong force splitting off the other two. Because of the rapid growth, all bumps and crevices in the curvature smooth out – which explains why the universe is so very little curved on large scales today.

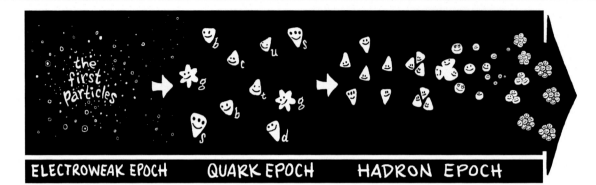

ELECTROWEAK EPOCH | QUARK EPOCH | HADRON EPOCH

10^{-32}–10^{-12} seconds

After reheating to temperature before inflation, temperature drops, until it reaches $\approx 3 \cdot 10^{15}$ K.

$\approx 10^{-29}$

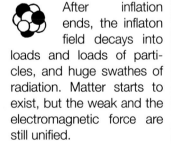

After inflation ends, the inflaton field decays into loads and loads of particles, and huge swathes of radiation. Matter starts to exist, but the weak and the electromagnetic force are still unified.

10^{-12}–10^{-6} seconds

$\approx 10^{15}$ K

$\approx 10^{-22}$

The universe is cooling enough so that the weak and the electromagnetic force split off. Now all fundamental forces take the form which they still have today. It is still too hot for the quarks to form protons and neutrons. So the universe is filled with a hot plasma of quarks, gluons, leptons, and all their antiparticles.

10^{-6} seconds–1 second

$\approx 10^{12}$ K

$\approx 10^{-15}$

The universe cools down enough so that quarks can form protons, neutrons and other hadrons. It is during this time that physicists expect that slightly more hadrons than anti-hadrons are created. The annihilation processes between matter and antimatter let mostly hadrons remain.

| **LEPTON EPOCH** | **PHOTON EPOCH** |

1 second–a few minutes

half an hour–380,000 years

$\approx 10^9$ K

$\approx 10^9$ K–3000 K

$\approx 10^{-10}$ or 0.0000000001

about 1/1090 at its end

 The universe is still filled with mostly radiation, as well as hadrons, and leptons and anti-leptons. Now the same thing that occurred with hadrons happens with the lighter leptons: there is a slight imbalance between the two, so that after mutual annihilation, only leptons (and of those, mostly electrons) remain in the universe, but they don't form atoms yet.

 The universe is filled with a plasma of atomic nuclei and electrons, very much like in a fusion reactor. In fact, a photon cannot fly for more than fractions of a second before it hits either a nucleus or an electron. It is only at the end of this period, which is called recombination, that the universe cools down enough so that nuclei and electrons combine to stable atoms. It is then, roughly 380,000 years after the Big Bang, that the universe becomes transparent – and photons can fly around freely. It's these earliest photons that we can still see in the cosmic microwave background.

"We built it to discover just what the universe was doing <u>before</u> the big bang to get such an amazing afterglow."

The Cosmic Microwave Background
The Oldest Photons in the Universe

When one looks up to the sky at night, one can see the stars shining, their light reaching us from the depths of the cosmos. If one takes out a large telescope which doesn't just detect visible light, but all kinds of electromagnetic radiation, one can see much more: Gamma rays from distant galaxies, X-rays coming from cosmic radiation, visible light from stars, galaxies and nebulae, and the deep infrared glow from dust and hydrogen which floats through our galaxy between the stars.

But there is another radiation, behind everything else, which has puzzled researchers when they first measured it, and which even today is not completely understood. When these signals were first picked up in 1964, it was first thought to be static, or a measurement error in the antennas.

It comes from every direction – even if you point your telescope towards the darkest point in the night sky, in between all the stars and galaxies, you will still see it. It can be picked up best with a radio telescope, and its wavelength (↗¹) is in the area of a few inches, which is why it has been dubbed cosmic microwave background, or CMB, in short.

The Cosmic Microwave Background – the Light Between the Stars

There are several things which distinguish the CMB from other radiation. First of all, as we have already mentioned, it comes from every direction with pretty much the same strength. So there is no single source, but that radiation rather fills the whole universe (there are about 400 photons of the CMB in every cubic meter).

But what is even more puzzling is that it is in thermodynamical equilibrium. What this means is that all photons that belong to the CMB have a very specific distribution of energy, as if they had all been interacting with each other, and exchanging energy back and forth. That is quite unusual for radiation. For example, the photons coming out of a laser do not have any temperature, because they all have precisely the same energy. Radiation coming from radioactive decay does have a certain energy distribution, but not a thermodynamical one: radioactive radiation doesn't have a temperature either.

But, for instance, the light particles coming from the sun do have a thermodynamic distribution of energy. And these light rays can be assigned a temperature – it is precisely the temperature of the sun, a couple thousand Fahrenheit. So what is the temperature of the CMB? Well, roughly 2.7 Kelvin, or about -455 degrees Fahrenheit. This is horribly cold!

As we have already hinted at, in order for many particles to be in thermodynamic equilibrium – to have a well-defined temperature – they need to have exchanged energy back and forth for a long time, until it has distributed roughly (but not precisely)

↗¹: *"Light" on page 7*

evenly. But photons do not do that. The only way to exchange energy is by bumping into each other, and photons cannot interact with each other. The only way for them to do this is indirectly: a photon bumps into an atom, and that atom bumps into another photon. That way energy can be transferred from the first to the second photon. This is what happens in the sun all the time, and it is the reason why the radiation which leaves the sun is in thermodynamic equilibrium.

In fact, this is the only way we know how this can happen. One simply needs those mediating atoms, or some other kind of matter! In other words, there is good reason to assume that the radiation from the CMB is some thermal radiation, emitted from something like a gas, or plasma, or whatever, with a certain temperature. But what should that be, and in particular, where should that be? The CMB is all around us, but where is that plasma?

An Afterglow from 14 Billion Years Ago

The short answer is: that plasma is us! Well, it was us – to be completely precise, the CMB is the afterglow of a time, when the whole universe was filled with a hot and dense plasma, which was pretty evenly distributed. That plasma was everywhere, and it was in thermodynamic equilibrium, and therefore had a certain temperature. Now since plasma consists of lots of positive and negative particles, the photons of that time could not fly for a second without bumping into a particle (photons just love to interact with charged particles). Therefore, all the photons in the universe were also in thermodynamic equilibrium with that plasma, and had the same temperature.

Because the universe expanded at that time (actually, it still does, ↗2,3), that plasma slowly cooled off, and so did the photons. That is, up until a certain time, when the universe was cooled down enough so that all the positive and negative particles combined to form the electrically neutral atoms (mostly hydrogen) that make up the matter of the universe. Photons interact with neutral atoms much less likely than with charged plasma, so suddenly the photons had no one to interact with any more. From one moment to the other – also called recombination – the universe became transparent for light particles. Since then, the photons from that time fill the universe, and make up the CMB. They are the weak afterglow from a time when the universe was a much less hospitable place.

The matter particles from that time have long since fallen out of thermodynamical equilibrium, and have taken on more interesting form, such as stars, nebulae, galaxies, planets, continents, trees and strawberry ice cream. But the photons from that time are still everywhere, and are still in thermodynamical equilibrium. The only thing which has happened to it is that it has cooled off. But that did not happen because it has given off its energy to somewhere else. Rather, it has cooled off because the universe has been expanding so much.

Calculations indicate that recombination happened about 377,000 years after the Big Bang – about 13.4 billion years ago! Since then, the size of the universe has increased by a factor of about 1100.

↗2: *"The Big Bang" on page 97*
↗3: *"Timeline of Our Universe" on page 100*

Now, for photons, the energy that they have is related to their wavelength (↗1). If space itself expands, so does their wavelength. Since higher wavelength means lower energy, the temperature of the CMB has decreased by a factor of 1100 since recombination. At the time when the universe became transparent, the plasma temperature was about 3000 Kelvin (about 5000 degrees Fahrenheit). Today, it only has a 1100th of that – about 2.7 Kelvin.

What the Oldest Light in the Universe Tells Us About Its History

One can rightfully say that the CMB consists of the oldest photons in the universe. All other photons have come into existence after recombination. And because the CMB has been relatively undisturbed since that time, it is a great relic for us to investigate. In fact, one can read many different things from the precise signatures within it. You see, the CMB is not precisely the same in every direction. There are tiny, miniscule temperature fluctuations up to 0.004 Kelvin. Although that does not sound like much, it tells us that the plasma that filled the universe, and later on formed the matter we consist of today, had the ever so slightest density fluctuations. Left alone, these fluctuations generically increase over time because of gravity (↗4). We can also compare this with what we know from gravity, and with the way ordinary matter is distributed in the universe today (↗5). This is one way to infer that there needs to be some kind of dark matter (↗6), for instance. Recently, there have been claims by the BICEP mission that one can read off imprints of the earliest gravitational waves (the left over shudder from the Big Bang, ↗7) in the polarization of the CMB. Unfortunately, as of the time of writing this article, it seems like the data was a result of imprecise measurements.

↗4: *"Birth of the Solar System" on page 59*
↗5: *"Large Scale Structure of the Universe" on page 109*
↗6: *"Dark Matter" on page 133*
↗7: *"The Big Bang" on page 97*
Image: ESA / Planck Collaboration

"How many rooms would you have to paper to begin to see an actual pattern?"

Large Scale Structure of the Universe
A Network Made of Stars

The universe is full of stars. But how are all the stars distributed? Are they all huddled together, like freezing people around a fire? Or are they sprinkled evenly, as raisins in bread? The short answer is: both! The longer answer is surprisingly complex.

First of all, you might know that our sun is only one of about a hundred billion stars in a galaxy, which we call the Milky Way. Most of the burning lights we see when we look at the night sky, are those.

It appears that most stars in the universe are in a galaxy – or rather, we only know of those, because single stars outside of the Milky Way are far too faint to see with our telescopes. So the next question is: how are the galaxies distributed? As it turns out, galaxies seem to huddle together in larger groups of tens to hundreds, called galaxy clusters. Many of these contain a mass of about 10^{14} times the mass of our sun, and are up to several million light years across.

The Milky Way is in a cluster called the local group, which we share with many well-known other galaxies, such as the Andromeda Nebula, the Pegasus galaxy, and the Large Magellanic Cloud.

Voids, Filaments and Superclusters

In the past years, several attempts have been made in order to understand how these galaxies and galaxy clusters arrange themselves. The image on page 111 shows the results from the "Two Micron All Sky Survey" – or 2MASS. Each point is a galaxy, and the color indicates the redshift, i.e. whether the galaxy is closer (more blue), or further away (more red) from us. As one can see, the galaxies and galaxy clusters are, in fact, also not evenly distributed. Rather, they arrange themselves in galaxy superclusters, where many galaxy clusters are clumped together. These superclusters can contain thousands of galaxies, and extend over millions of light years. Different superclusters are connected by thin, fiber-like structures called filaments, on which galaxies arrange themselves like beads on a string. Also, there are large, unoccupied regions called voids, which can also be millions of light years across, and contain none, or at least almost no galaxies.

BLOBS   filaments   VOIDS

This network-like structure seems to go on forever – at least it fills the whole observable universe, which has a diameter of about 90 billion light years. We cannot see regions further away, because the light from beyond did not yet have enough time to reach us since the Big Bang. And on these large scales, the weave of filaments, superclusters and voids is so fine, that the average density of galaxies is more or less the same everywhere. So on very large scales, the matter in the universe is, in fact, very evenly distributed.

Ripples in the Cosmos: It's What Gave the Universe Its Structure

Where does this distribution of galaxies, forming superclusters, filaments and voids, come from? The answer is simple: gravity! In fact, we have a quite good idea of the processes which led to the large scale structure of the universe since the Big Bang (↗[1]). Let's go back to the process called recombination, when the universe was about 300,000 years old (↗[2]). At this time, matter and radiation separated from each other, and the universe became transparent for light. The light from that time can still be seen today as the radiation of the Cosmic Microwave Background (CMB), and it is quite homogenous. So also all the atoms (mostly hydrogen), which were in thermodynamic equilibrium with the radiation before recombination, must have been very evenly distributed over the entire universe. But as we know, there are tiny temperature fluctuations in the CMB today, so there must have been equally tiny density fluctuations in the matter back then.

Because of gravity, every piece of matter attracts all other matter. If it were perfectly evenly distributed in the universe, the overall force on each particle would cancel out. The matter would always stay completely homogenous. But if there are even the smallest ripples in the matter density, these inhomogeneities grow over time. Regions which are slightly denser than average pull a little bit stronger on the other matter, attracting more and more of it, and getting ever more dense. Over time, the matter of the universe clumps together. This happens on

relatively small scales, leading to the formations of stars and solar systems (↗[3]), as well as on the largest scales – there it just takes much longer.

Small Structures Make up Large Ones

But if that is true, why isn't the whole matter organized in clumps? It seems to be, at least on relatively small scales: There are stars, which can be seen as pretty round clumps of matter. Many stars are packed together in galaxies, which are more or less spherical, depending on how fast they rotate. Several galaxies are arranged in galaxy clusters, which are also pretty blob-like. But if we look at larger and larger scales, we start to see the filaments and voids. Why don't superclusters form supersuperclusters, and those supersupersuperclusters, and so on? The answer lies in something called hierarchical structure formation.

To understand hierarchical structure formation, one needs to have a closer look at how a large blob of matter actually collapses under its own gravitational attraction. Look at an ellipsoid – this is like a spherical ball, just that it has three different diameters, one from left to right called x, one from front to back called y, and one from top to bottom called z. Let us look at the example where x is larger than y, which in turn is larger than z. So the ellipsoid looks a bit like a thickened pancake. If the matter density is the same everywhere in this pancake, how does it contract under its own weight?

One could assume that it just contracts by shrinking in a regular kind of way, but that is not the case.

↗[1]: *"The Big Bang" on page 97*
↗[2]: *"The Cosmic Microwave Background" on page 105*
↗[3]: *"Birth of the Solar System" on page 59*

Rather what happens is this: First, the smallest diameter collapses, so z becomes really small, until we have a quite flat pancake. Then the next larger diameter, y, becomes smaller, until the pancake contracts into a slightly thick line, or

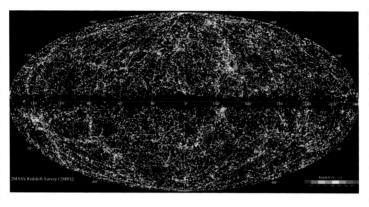

a straightened earthworm. The largest diameter x contracts latest, which is when the earthworm contracts to form a round blob.

This is why the universe looks the way it does: Just after recombination, the matter of the universe had just the tiniest density fluctuations, which have amplified over time. Not in a regular way, but via hierarchical structure formation. Over very large distances, the contraction takes much, much longer, because the gravitational pull has to reach further. So on relatively small scales, the matter of the universe has coalesced into the final, blob-like form, like stars or galaxies. These are relatively stable and don't contract any further. On large scales, however, the matter did not have enough time to arrange into blobs, so here we are still on the intermediate stages, where we have the long filaments (the earthworms), or the flat pancakes, which surround the voids. On the largest scales, the irregularities in the densities have not yet affected the overall matter distribution, so in the universe as a whole, matter is still distributed quite evenly.

Numerical Simulations since the Big Bang

By the way: We do not only have this qualitative picture of how the large scale structure of the universe formed. Nowadays, we also have quite extensive numerical calculations to back up these ideas. Take for instance the Bolshoi-simulation: Here are three instances of a simulation of the matter distribution in a box with 1 billion light years of edge length, taken 500 million, 6 billion, and 13.8 billion years – in other words today – after the Big Bang. The last simulation image looks pretty much like the actual universe today.

Image (upper): 2MASS, T.H. Jarrett
Image (lower): Kristin Riebe, MultiDark project

"We've finally discovered an unclassified galaxy formation that our shareholders can appreciate."

Galaxy Types
Looks Is Everything

The universe is full of stars, our sun being only one of literally uncountably many other balls of fire that light up the eternal blackness of space. These stars, however, are not distributed evenly. Instead, almost all of them are grouped together in galaxies (↗1). Each galaxy itself is a collection of hundreds of billions of stars, and so far we estimate about a hundred billion galaxies in the universe. At least in the part which we can see.

The Milky Way: a Spiral Galaxy

Our galaxy is called the Milky Way, because on a clear night, one can see it as milky band of stars that spans across the sky. The true beauty of the Milky Way, however, reveals itself only to those who could observe it from above: having a big, bright bulge in the middle, two bars extend to the outer rims, ending in four large spiral arms, called "Outer", "Perseus", "Sagittarius", and "Centaurus". Because the Milky Way, like most galaxies, rotates around itself, these arms coil around it like tentacles. There are also two minor arms, which do not extend quite as far. Our sun and solar system are in one of these, called "Orionis".

This is a view we will most likely never be able to witness ourselves. The furthest cameras we have are in the Voyager probes, which have barely just left the solar system (↗2).

They can see the Milky Way from above as much as someone who had just visited the neighboring street could see the shape of the American continent.

So how do we know our galaxy actually looks like that, when nobody has ever seen it from that angle? Well, for once there are several good ways to estimate the distance to other stars in the Milky Way, so even if we sit directly inside it, we can get a pretty good idea of our galaxy's shape. At least of the part which is close to us. On the other hand, we can compare it to other galaxies, which we can see from far away. Although there are billions of other galaxies in the universe, most of them fall into one of only a few different categories.

The Hubble Sequence: Elliptical, Lenticular, or Spiral?

It was Edwin Hubble who devised one of the first schemes, by which different galaxy types can be identified. This "Hubble morphological classification" distinguishes between three major types: the elliptical galaxies, the spiral galaxies, and the lenticular galaxies.

If a galaxy is elliptical, it basically means that it is a round blob, without much of an internal structure. The stars in elliptical galaxies are distributed evenly throughout it. They appear as smooth, feature-

---

↗1: *"Large Scale Structure of the Universe" on page 109*
↗2: *"The Voyager Probes" on page 55*

less blobs. Their class designation begins with an "E",
followed by a number from
0 to 7. This number signifies how elliptical they appear to us. So an E0 type
galaxy appears completely
round, like a circle, while an
E7 type is very elliptical. This
number actually signifies
how the galaxy is oriented
to us: All elliptical galaxies
are flat pancakes, it is just
that we look at E0 type galaxies from above, while we
look at E7 type galaxies edge-
on. The most common elliptical type
one finds in the universe is E3.

Originally, Hubble thought that being elliptical was
an early stage of galaxy evolution. So a galaxy would
be elliptical, before it would evolve into
the other types (see below). But
the opposite seems to be
true: it has been found that
the stars in elliptical galaxies are usually quite old,
and that there is very little
interstellar matter in them.
Since interstellar matter is
the main driving force of star
formation (↗³), there are not
many new stars forming in
them. It is assumed that elliptical galaxies are the oldest ones, probably the result
of the merger of two galaxies.
When the Milky Way and the An-

dromeda galaxy merge in 4–5
billion years time, the result will
probably also be an elliptic galaxy.

The lenticular galaxies, just as
the elliptical galaxies, have a
bulge of stars, but additional
to that they also have a thin
disc around it, in which some
of its stars can be found.
They could have also been
called the "sunny side up"
type galaxies!

Much like the elliptical galaxies,
they have either used up or lost most
of their interstellar matter. So not much star formation is going on in them neither. Some researchers
suggest that they are an intermediate stage between the spiral galaxies and the elliptic ones. If the
disc is hard to spot in a telescope,
one might easily mistake a lenticular galaxy for an elliptic one,
and indeed there are some
examples where there is an
ongoing debate of which category they should be in. The
lenticular galaxies are desig-
nated as S0.

The last category is that of
spiral galaxies. Here two important sub-cases have to
be distinguished: Does the
galaxy have bars or not? This
is not to signify whether they
have places to hang out after

↗³: *"Birth of the Solar System" on page 59*

work, but rather whether there are two or more handlebars emerging from the central bulge.

Spiral galaxies have – as their name suggests – large spiral arms that emerge from their center, wrapping around them. Andromeda is a spiral galaxy, as is our Milky Way. In fact, these spiral arms have shaped our image of what a galaxy looks like so much, that it is often hard to imagine that there are other types of galaxies in the universe as well.

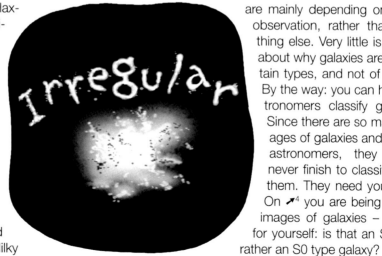

The spiral galaxies are distinguished further by how tightly the arms are wrapped around the center. The ones without bars have just one small letter following them: Sa galaxies have many, very tight spirals, which can make it hard to distinguish them from S0 galaxies. Sb and Sc have fewer and more spread out spiral arms. Sd galaxies have very prominent arms, which are much brighter than even the bulge.

And the ones with bars? They get another "B" to their name, making them into SBa, SBb, galaxies, and so on. By the way: our Milky Way is most likely a SBc type galaxy!

Whenever people are inventing some kind of classification, to bring order to the chaos that nature presents to us, there is always one special category: the "doesn't fit anywhere else" category! That one also exists in the galaxy morphology, and the galaxies which fall into them are called irregular. In fact, they are indeed further subdivided – there are galaxies of class irregular 1 and irregular 2, depending on whether they are completely asymmetrical and lack a central bulge (Irr 1), or have at least some smooth features (Irr 2).

As with all these classifications, they are mainly depending on visual observation, rather than anything else. Very little is known about why galaxies are of certain types, and not of others. By the way: you can help astronomers classify galaxies! Since there are so many images of galaxies and so few astronomers, they would never finish to classify all of them. They need your help! On ↗4 you are being shown images of galaxies – decide for yourself: is that an SBa, or rather an S0 type galaxy?

↗4: http:// www.galaxyzoo.org

Relative Space and Time
Why you Can't Make Light Faster by Pushing it

Sometimes one can hear the phrase "everything is relative", in order to describe the theory of relativity by Albert Einstein. This is, of course wrong: If everything was relative, then there'd be nothing anything could be relative to. Rather, the Theory of Relativity claims that some things we believed to be absolute, are in fact relative – like space, and even more importantly time. On the other hand, it also requires that some things we had thought were relative, are in fact absolute – such as the speed of light.

Newton's Space and Time Were Absolute!

In 1905, when Einstein published his relativity theory, this came as quite a surprise. For about three hundred years, the world view of Sir Isaac Newton had dominated the way everyone thought about space, time and matter. It went like this: Space and time were absolute, fixed and unchangeable. This meant, for instance, that space could be thought of as the eternal "stage", on which all matter "played", i. e. where particles moved and interacted with each other. In particular, space was unlike, and independent of matter.

Even more fundamental was the absoluteness of time. Time was thought to always flow with the same speed (as in "one second per second"), and to pass at the same rate everywhere, independently from whomever experienced it. Crucially, one could talk about events at different places happening simultaneously. Two different events could happen at different places, but "at the same time".

All of this makes intuitive sense today, as it did in the 17th century. After all, we have the feeling that "What is happening on the other side of the earth right now?" is a sensible question to ask, even if it might involve some effort to find the answer. Unfortunately, that is not how the universe actually operates: strictly speaking, the question does not even make sense! And the only reason, it is so hard for us to notice this fact, is that light is so incredibly fast compared to everything else on the Earth – in one second, a light ray can travel 7.5 times around the planet!

For Us, Speed Seems Relative – but for Light It Actually Isn't!

How does that work, precisely? Well, it starts with the observation that the speed of light is constant for all observers. This is actually quite different from what we are generally used to! In our everyday life, speeds are relative: Imagine you sit in a fast car on the highway. From your point of view, another slightly faster car overtaking you appears to be crawling past you at a painstakingly slow pace. For someone who watches it from the side of the street however, that car is actually really fast. So the apparent speed of the second car is, in fact, dependent on the observer.

Also, that is why a javelin thrower takes a "run-up" to gain some momentum before she throws the javelin. From her point of view, the speed of the javelin leaving her hand is the usual throwing speed. From the point of view of the referee however, it is the throwing speed plus the running speed of the athlete. And it is the referee who judges how far the javelin has been thrown, so, of course, the athlete wants to maximize the javelin speed for him.

All of This Is Different for Light!

For the sake of the argument, assume that instead of throwing a javelin, the athlete would switch on a flashlight while running (those would be kind of boring Olympics, but bear with us for a moment). Then, the speed of the light shining from the flashlight would actually be the same for the referee and the athlete. Even though one is running over the grass, and the other is standing still. If they had devices to measure the speed of the light particles (called photons, ↗[1]) from the flashlight, they would both measure exactly the same value: about 187.500 miles per second.

How can that be? Isn't one of the two making a mistake? Isn't throwing a javelin the same as throwing photons? Surely, if the latter are being shot out of the flashlight, then one should be able to make them even faster by moving the flashlight along as well, no?

No, and here's why: The speed of light is absolute, because the passage of time is relative! More specifically, the notion of simultaneity is different for the referee and the athlete. Now here is a quite convoluted way of measuring the speed of light, which illustrates our point: Assume the referee is some distance (say, 100 yards) away from the athlete at that

moment when she switches on the flashlight. She holds it into the referee's direction, who carries a (very, very precise) stopwatch. Because the athlete wants to be thorough, she also carries one. They agree on the following: The athlete is running towards the referee, and when passing the 100 yard line, she switches on the flashlight, and starts her stopwatch. At the same time, the referee also starts his. As soon as the photons from the flashlight arrive at the referee, he stops his stopwatch, and at the same time the athlete also stops hers.

Usually, what one would expect is the following: To determine the speed of the light particles, they need to divide the traveled distance by the time they needed to travel it. Both start and stop their watches at the same time, so they measure the same time difference. But for the referee the light has to travel 100 yards, where from the point of view of the athlete, the light has to travel a bit less than 100 yards. This is because she is running towards the referee, so from her point of view, the referee is coming towards her, effectively shortening the distance a bit. So the speed of light from the referee's point of view should be a bit larger than from the athlete's point of view.

As we have already said, this is not what happens! For things that are very slow compared to the speed of light, it is a very good approximation, but as soon as things get really fast, one has to take the relativistic nature of time and space into account!

Simultaneity Confounded!

This is what actually happens: The referee and athlete will slightly disagree on when to exactly start

↗[1]: *"Light" on page 7*

and stop their stopwatches! They will simply not be able to do it "at the same time", because these particular moments will be different for them. From the point of view of the referee, both start their watches at the same time, but the athlete will stop hers prematurely. So for her, the athlete will have measured a shorter distance, but also for a shorter time.

From the point of view of the athlete however, both will stop their watches at the same time, but the referee will have started his watch too early. So from her point of view, the referee measured a longer distance, but also for a longer time. If they later on meet to compare their numbers, they will find that the ratio will yield the same result: The speed of light for both will be the same, because the moments of simultaneity were different!

It is important to note that neither of them is wrong and/or right. This is the crux of the theory of relativity: The passage of time depends on the observer! More specifically: it depends on the way observers move relative to each other. By the way: in real life, the effect will only be a few fractions of a second – far too small to actually notice with ordinary stopwatches. But this is because we are dealing with the notion of two things happening at the same time, being only 100 yards apart (in our case "athlete switches on flash light" and "referee starts his stop watch"). If this distance increases, so does the time difference. So it really makes no sense to ask "What is happening right now in the Andromeda galaxy?" The Andromeda galaxy is XYZ away, so two people, slowly walking past each other on the street, would disagree about when "right now in the Andromeda galaxy" exactly is, by about a few million years!

This effect has some very peculiar consequences. The first, and most well-known, is the fact that nothing can go faster than light. The speed of light is the magical barrier that cannot be crossed (see ↗², though). In particular, you cannot make light itself any faster by pushing it along. Conversely, you will not be able to slow it down – it will simply always have the same speed, no matter how fast you yourself are.

Relativity also affects all precise time and length measurements. For example, it appears that fast moving clocks run slower. This is called time dilation, and it is quite important for a lot of things, most notably the GPS navigational systems (also, if you fancy, google "twin paradox"). The GPS satellites orbit the Earth twice per day, and rely on extremely precise clocks. So precise, in fact, that effects from relativity have to be taken into account, otherwise it would deliver the wrong position: The error would be even growing over time, by a few yards per day.

So next time you use your cell phone to locate the nearest Starbucks, remember: you would not have found it without the theory of relativity!

↗²: *"Warp Drive" on page 295*

The GENERAL RELATIVITY AMUSEMENT PARK

The Theory of General Relativity
Curved Space and Warped Time

In 1905, Albert Einstein's theory of relativity (↗¹) shook the world view of physicists. With his equations, Einstein had overthrown many principles that Sir Isaac Newton had formulated about two centuries earlier. These principles – in particular the ones about absolute space and time – had been held very dear by everyone, from physicists to philosophers. So people adapted only slowly to the new world view.

Gravity Revised

But Einstein did not stop there. He was thinking about the other laws that Newton had stated: in particular the one about gravity. Newton's law of gravity had been around for centuries, and it had explained the motion of planets, moons, stars, and apples falling onto one's heads. For everything, from calculating the position of Jupiter in the sky to the trajectory of ballistic projectiles, Newton's law of gravity delivered the correct answers, to an astounding degree of accuracy. Einstein was wondering: How would this law of gravity look like, when taking relativity into account? Also, the laws for the electromagnetic force, formulated by James Clerk Maxwell a few decades earlier, fit very well into the scheme of relativity. So surely something similar should also be possible for gravity?

Surprisingly, it turned out to be quite difficult to reconcile gravity with relativity. One reason was that, in Newton's theory, the gravitational force acted instantaneously over large distances: If the sun were to suddenly vanish, we would feel the vanishing of the sun's gravitational pull instantly, claimed Newton's laws. But how could this fit together with relativity, in which no influence could travel faster than the speed of light (and the notion of "instantly" was highly problematic anyway)?

It didn't. In fact, to incorporate gravity into the framework of relativity, Einstein had to radically rethink how space and time worked – again!

His initial reasoning went like this: All bodies fall the same, no matter how heavy they are. So the gravitational pull at a certain point should be no ominous force, but actually a property of space (and time) at that place. It took him quite some time (and help from his wife) to work out the details, but ten years later, in 1915, he published his theory of "general relativity". His previous relativity theory from 1905 was henceforth dubbed "special relativity". This was because it only treated the special case where the gravitational attraction between masses is not important, and can be ignored.

Gravitation or Acceleration

The theory rests on the so-called equivalence principle. This principle states that one cannot distinguish between feeling a gravitational force, and being in an accelerated reference frame. So, if you were in a space shuttle, standing on the floor and

↗¹: *"Relative Space and Time" on page 117*

feeling the gravitational pull downwards, you could not decide whether the shuttle was either standing still on the Earth, or it was somewhere in space, but being accelerated upwards. You would have to look outside the window to decide this, the force you feel would be exactly the same. It would not matter if the origin of that force were either the gravitational attraction of the Earth, or you being pressed downwards by the shuttle being pulled upwards.

This also works for the absence of gravity: If you feel weightless, it can either be because you are somewhere in space, and are therefore floating around (hopefully in a space suit), or you are on Earth, but currently falling downwards. Both cases will feel precisely the same to you. Well, at least until you hit the ground.

two parallel lines only meet at infinity, the sum of angles in a triangle always add up to 180 degrees, and time passes at the same speed in different places.

In general relativity, the situation is different. Einstein realized that if heavy masses are present, space-time is actually not flat, but curved. As a lead weight makes a dent in a rubber sheet, so do heavy masses make a dent in space-time. Everything that has a mass, or even just an energy density, such as a light ray, distorts space and time around it.

On the other hand, objects that move through the curved space-time appear as if they are influenced by gravitational attraction. However, there is no force acting at a distance between objects. Rather, it is the curved geometry of space and time which forces them to come closer to each other.

In special relativity, the three directions of space and the one direction of time were recognized as not being two different concepts, but as two inseparable parts of a whole: four-dimensional space-time. In its 1905 version, space-time is always flat, like a piece of (four-dimensional) paper. This means that

Here is a simplified version of what is happening, which only involves the curvature of space, rather than time: Imagine two ships in the Atlantic ocean,

both on the equator, one being a few miles east of the other. Both ships now head north. They look at their compasses, turn their ships directly towards the North Pole, and start their engines. At first, they will keep their constant distance, but the further north they get, the closer they will come to each other. If Greenland were not in the way, they would meet at the North Pole. It will look like the two ships attracted each other, even though both ship captains will swear that they were driving in a straight line! This is because the surface of the Earth is not flat, but curved. In a similar way, the curvature of space and time will cause massive objects to gravitationally attract each other.

Gravity's Effect on Time

There are some really weird consequences of general relativity: One of the strangest is gravitational time dilation (↗1). Near very heavy masses, time passes slower! On Earth, this effect is quite small, but actually noticeable if you have very precise clocks. So when one year passes down here on Earth, in space far away from Earth one year and about 0.02 seconds would pass. Not very much, you might say. But crucial for any space mission, which need to be timed to an incredibly precise accuracy! In 0.02 seconds, a point on Earth's equator – such as a landing strip – moves about 10 yards due to Earth's rotation. It's embarrassing if one botches the landing with a space shuttle, simply because one forgot to take general relativity into account.

In fact, Newton's equations are sufficient for most of the phenomena in our solar system. One exception is the fact that the perihelion (the point of nearest approach to the sun) of Mercury's orbit is changing slightly over the years. Much of that change could be explained by the gravitational pull of the other planets, but there remained a small part which could only be explained by general relativity. So this was one of the first confirmations of Einstein's new theory. Observing gravitational lensing (↗2) for the first time was another.

One of the most drastic consequences, however, is the fact that the universe cannot be stationary and unchangeable, but either has to contract or expand (↗3). That is, unless you add an artificial fudge factor to Einstein's equations, called "cosmological constant". The observation of the American astronomer Edwin Hubble in 1929, that the universe is, in fact, expanding, decided this question thoroughly. At that time, Einstein revoked the cosmological constant from his equations, calling its introduction his "greatest blunder". After 1990, detailed observations of astronomers suggested that the cosmological constant might in fact be added to the equations, and that the expansion of the universe is even accelerating (↗4)!

↗2: *"Gravitational Lensing" on page 129*
↗3: *"The Big Bang" on page 97*
↗4: *"Dark Energy" on page 137*

Curved Space Time
Getting the Right Angle

More than a hundred years ago, Albert Einstein published the theory of general relativity (↗[1]). The term "general" here refers to the fact that you can describe more general situations than those which the theory of special relativity (↗[2]) is able to treat.

Special relativity describes how single – not too heavy – bodies move through empty space. In particular, it asserts that nothing can move faster than light. General relativity, however, is also able to describe how different bodies attract each other via the force of gravity. It was Einstein's great achievement to work out that the force of gravity cannot be just slapped on to special relativity – it needed to be completely redone! In particular the concept of an unchanging space and absolute time had to be scrapped. And this after more than 250 years of Newton, who chiseled these notions into the subconsciousness of generations of scientists!

Curved Space-Time: How Does It Work?

General relativity states that a lot of matter – such as a star, for instance – curves space and time. Ultimately, if another object – such as a planet – moves in this curved space-time, its movement path becomes distorted, in such a way that it looks like there is a gravitational attraction between star and planet. But what does this curvature actually mean?

How does curved space "look like"? How would we realize that we were actually moving in curved space? And, even more mysterious, what is curved time? All of this is actually not so easy to describe. Incidentally, the mathematics of curved spaces had to be worked out completely shortly before Einstein actually needed the formulas for his theory. The German mathematician Carl Friedrich Gauss had figured out how to describe curved surfaces, such as the surface of the Earth (which he needed, because he also worked as a land surveyor). Later, his student Bernhard Riemann had completely figured out the mathematics of curved spaces of arbitrary dimension (not just two-dimensional surfaces).

Two-Dimensional Curvature

Let us first have a look at two-dimensional surfaces. The geometry of surfaces, in particular that of the plane, had already been of great interest to the Greek: they regarded the science of lines, angles, circles and triangles as a chief discipline among mathematicians. One of them, Euclid, asked himself: "What is the plane, after all?" He came up with his famous postulates, the first four of which determine the foundation of what it means to be a plane: 1.) There are points in the plane, and you can join two points by a straight line segment. 2.) Any line segment can be extended up to infinity, to become a straight line. 3.) For any straight line segment, one can draw a circle having one of its endpoints in the

↗[1]: *"The Theory of General Relativity" on page 121*
↗[2]: *"Relative Space and Time" on page 117*

center, and the other endpoint lying on the circle. 4.) All right angles are the same everywhere (the Greek used the word congruent for this).

Euclid's Axioms

As it turns out, you can do a lot of geometry already, just with these four rules. There was one thing, though: The well-known fact that the sum of the three angles in a triangle was 180° – that he could not prove, just using his four rules. He needed to add a fifth one: 5.) If two lines are drawn which intersect a third in such a way that the sum of the inner angles is less than two right angles, then the two lines inevitably must intersect each other on that side if extended far enough.

Doesn't quite have the same ring to it, does it? It seemed quite convoluted to Euclid, and so he – and generations of mathematicians after him – tried to work without it, or just show that it was actually a consequence of the first four. Then one could have left it out. But alas, they all failed. Today we know the reasons for that: there actually are spaces, in which the first four rules can be applied, but the fifth one can not! These are also called "non-Euclidean spaces", and their difference to the plane is that the sum of the three angles in a triangle is not necessarily 180° (or "two right angles", as Euclid would have put it). The reason is that these spaces are curved.

Triangles and Their Angles: an Indicator of Curvature

In fact, this is the essence of curvature, in general: triangles' angles don't add up to 180°. Here is a way in which you can prove that the surface of the Earth is curved: If you are in New York, then consid-

er the directions in which, say, Rome and London lie. Write down the angle between these two directions. Then travel to Rome (or ask someone living there) to do the same thing with the two directions in which New York and London lie. Similarly with the two directions to Rome and New York, from the point of view of someone in London. If you sum up the three angles that you wrote down that way, you will get around 250°! This is because the triangle with the three corners at New York, London and Rome lies on the surface of the Earth – which is curved! More precisely, it is positively curved, which is why we get a lot more than 180°. There are also negatively curved spaces, where the sum is actually less than that.

Now, the surface of the Earth is two-dimensional. This means that one needs two numbers to describe a point on it – in this case longitude and latitude. Our space-time, on the other hand, is four-dimensional: one needs four numbers to describe an event. For instance, one could use longitude, latitude and height over ground (three numbers) to say where, and time on the clock (one number) to say when something happened.

Curvature in Four Dimensions: so Many Triangles!

To describe the curvature of four-dimensional space-time is much more complicated than to just use the words "positively curved" or "negatively curved". This is because there are many different ways in which one can place a (two-dimensional) triangle in (four-dimensional) space-time. There are just more directions you can orient it. That is why the formulas in higher-dimensional geometry, such as general relativity, tend to be quite cumbersome.

There is a question which is asked every time, when one tells the story of general relativity: "If space-time is curved, what is it curved in?" The two-dimensional surface of the Earth is positively curved, but Earth itself sits in three-dimensional space. And if you are allowed to use that, there seems to be no curvature: If you do not take the angle between the directions you'd have to travel by ship in order to reach London and Rome, but take the angle between the directions you'd have to drill through the Earth in order to turn up in London and Rome, you would end up with a triangle in three-dimensional space. So the three angles measured that way would actually (very nearly) add up to 180°. So this curvature of the surface would only be "accidental", because the Earth is a part of the three-dimensional space. If four-dimensional space-time appears to be curved, what is it curved in?

What Is Space-Time Curved in?

Well, nothing! There is one important point about curvature: it comes in two varieties. One is called the extrinsic curvature. A space has extrinsic curvature, because it is embedded in some larger space, just as the surface of the Earth is embedded in three-dimensional space. The other is called intrinsic curvature, which is a property of the geometry of the space itself. Unfortunately, two-dimensional surfaces have the special property that extrinsic and intrinsic curvature are always precisely the same: there is no intrinsic curvature which you could "generate" by somehow putting the surface into three-dimensional space. That is why it is diffi-

cult to explain the difference between the two with, say, the surface of the Earth. But there is one, and the four-dimensional curvature is, indeed, intrinsic.

Intrinsic and Extrinsic Curvature

One way of imagining the difference between intrinsic and extrinsic curvature is by picturing the circle on a flat plane: It is one-dimensional, and all one-dimensional objects cannot have intrinsic curvature (basically because you cannot put triangles in them, to compute sums of angles). In particular, a one-dimensional observer moving inside our circle will only know that it can move forwards or backwards. It will not be able to realize that it runs round in circles, because is has no concept of "sideways". So there is no intrinsic curvature. But we, as observers from the outside, can see that the circle is bent – so it has extrinsic curvature.

The situation is similar with three-dimensional space, and four-dimensional space-time, by the way. Three-dimensional space (the universe), at least on average over large distances, is flat (↗3). This means that we, as inhabitants of the three-dimensional world, do not see any curvature. A triangle with end-points Earth, Andromeda galaxy, and the Large Magellanic Cloud, will have angles summing up to 180°. But the whole of space-time is, indeed, curved intrinsically. The three-dimensional universe at a certain time (after the Big Bang, ↗4), again, is part of four-dimensional space-time, which is why the universe has extrinsic curvature – we see this as the fact that the universe is expanding over time.

↗3: *"Dark Energy" on page 137*
↗4: *"The Big Bang" on page 97*

"We know the universe's <u>age.</u> Gravitational lensing will help us determine its <u>weight</u>. If the universe is female she's going to give us a hard time."

Gravitational Lensing
Mirages in the Night Sky

Seeing is believing – but often, what we see is not really there. Or it is there, but at a different place than the one we think it is at.

We see things by observing the light they emit, which reaches our eyes. When we try to infer where the observed object is exactly, our brain instinctively draws a direct line from our eye into the direction of where the light ray is coming from. But sometimes light rays do not travel on straight paths. That is when our brain is fooled, and we see things that are not where we think they are.

The bending of light rays occurs often in nature. For instance, light is refracted when passing from water to air, which is why it is so difficult to catch a fish with your own hands: it is not where you think it is. Also, light rays are distorted slightly when passing through hot air. This is why you see the air flicker over a hot road, or why some people have been led astray in the desert by illusions of water which wasn't there. But light bending is also used for good: glasses for instance are used to distort light in such a way that defects of the eye are counteracted, and make people see things sharply they otherwise would not. Also, microscopes can magnify the smallest bacteria, for us to look at, by clever bending of light rays.

There is one other very important instance, where light does not necessarily follow straight lines: When it travels long distances through the universe.

Light in the Universe Does Not Always Travel along Straight Lines

When we observe the universe around us through telescopes and observatories, we rely on light that reaches the earth through the interstellar space. But this light rarely follows completely straight lines. The reason for this is gravity: it is the gravitational curving of the geometry of space itself (↗[1]), which leads to the distortion of the path light takes, when it travels through the universe. So, also in space, one can sometimes see things which are not exactly where we think they are. This effect is called gravitational lensing, and it is of crucial importance for the astrophysicists who measure and observe the cosmos.

Most of the time, it's heavy masses which curve space, and therefore bend light rays. As a rule of thumb, the heavier the object, the more the light is bent. A relatively small object, like, say, the moon, is hardly heavy enough to influence the light around it. But a star is already so massive that it slightly changes the paths of light rays which pass it nearby.

↗[1]: *"The Theory of General Relativity" on page 121*

Take the sun, for instance: As stars go, it is one of the lighter ones, so it doesn't distort light rays too much. In fact, in 1919, Sir Arthur Eddington looked very closely at the sun during a solar eclipse, and found that the stars close to it appeared at different locations than they should have – by the exact amount predicted by general relativity. This made Albert Einstein famous over night: newspaper headlines reading "Space Is Warped!" went around the world, speaking of the first confirmed observation of gravitational lensing.

Gravitational Lensing Is Everywhere

Today, the effect of gravitational lensing can be seen everywhere in the universe, and it is a very convenient tool for astronomers to calculate the mass and/or distance of far away stars or galaxies. A very prominent example is the so-called Einstein Cross: If you look at a certain point in the constellation of Pegasus with a good telescope, you can see five light points forming a cross. The center of it is a galaxy, about 500 million light years away, which carries the poetic name UZC J224030.2+032131. The other four dots of lights surrounding it are actu-

ally multiple images of the same object: the quasar QSO 2237+0305. The quasar (actually an incredibly bright galaxy, ↗²) is about 10 billion light years away, and is located behind UZC J224030.2+032131, from the point of view of the earth.

The gravitational lenses found in the universe are not quite like optical lenses. There are a few important differences. Firstly, light rays which pass an optical lens close to its center are bent much less than rays which pass it further on the outer rim. With gravitational lenses this is the other way round: the closer the light passes the object, the more its path is changed. That is why, unlike optical lenses, gravitational lenses have no single focal point. Rather, they often have a focal line. So we do see the image of distant objects mostly distorted, rather than magnified.

Secondly, gravitational lenses are often not perfectly spherical, but are irregularly shaped objects, like galaxies. Therefore, also their gravitational field is irregular, and the precise way in which light is refracted is quite complicated. It can lead to multiple images, as in the case with the Einstein Cross. Only if the lensing object is near-

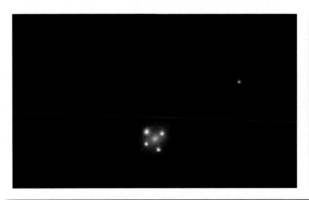

↗²: *"Galaxy Types" on page 113*
Images: ESA/Hubble & NASA

ly round, can one get to so-called Einstein rings: The image of an object behind the lens is distorted into a ring-like shape, which appears to eclipse the gravitational lens. This is, for instance, the case with the galaxy LRG (for luminous red galaxy) 3-757, shown in the right image on the previous page.

As already mentioned, the bending of light is stronger, the closer the light ray passes the object. Theoretically, this makes black holes (↗3) into great gravitational lenses. They can be as heavy as ordinary stars, but are much, much smaller, compared to their size. In fact, close to the event horizon of a black hole the gravity becomes so strong that a light ray passing nearby can orbit the black hole several times before traveling further. This is why, with an ideal black hole (that is, one that is not surrounded by lots of other matter which would obscure the view), one could even see several Einstein rings. This is certainly something that is completely out of the question with usual optical lenses.

Gravitational Lensing: a Way of Weighing the Universe

The effect of gravitational lensing is incredibly important for anyone trying to make sense of the images that are being shown by telescopes, no matter whether we talk about the huge dish arrays in the Peruvian highlands, or the ones mounted on satellites, such as HUBBLE. Many of the images we see are cosmic mirages.

But this effect is not simply a nuisance for those who want to draw accurate star charts. It can also be extremely helpful in weighing the universe. Light of stars and galaxies is distorted by gravitational lensing effects in a very characteristic way, which can be easily recognized. Also, from the red shift of this light one has a good estimate as to how far away it began its trip through the universe (↗4). With these two pieces of information, one can get a fairly good guess as to how much mass lies between the origin of the light ray and Earth. This is like trying to guess the temperature of a hot street by looking at the ripples in the air are above it – not easy, but definitely possible!

This method has been used, not just to calculate the mass of single galaxies, but also to get some information of the matter distribution within the whole universe. And then one can compare it with the total mass we get by counting all the stars, galaxies, nebulae and dust clouds we can observe anyway.

By the way, the result is quite surprising: The two numbers do not agree at all! There is much more gravity than there is matter in the universe. Well, at least than there is visible matter. Most of the mass in the universe, the presence of which we can infer by looking at the lensing effects on light, must be invisible (and with this we mean, properly invisible, not just hard to see because it is badly lit). This revelation actually fits well with lots of other independent observations, which would be explained by a new form of dark matter. We will come back to this in a separate article (↗5).

↗3: *"Black Holes" on page 91*
↗4: *"The Doppler Shift" on page 15*
↗5: *"Dark Matter" on page 133*

"You have to admit -- the black pudding was good."

Dark Matter
More than Meets the Eye

How can you tell something is there when you can't see it? Usually there are indications: tracks in the snow tell you that an animal must have passed through. The smell of bacon in the morning promises breakfast deliciousness, even if you are still in bed and can't see it yet. And if you look outside and see splashing puddles of water in the streets, you know it is raining, even if the falling droplets of water themselves are too small to make out. This is a bit how it is with dark matter: One cannot see it directly, but there are several indications that something must be there in the universe, even if it is invisible.

Missing Matter: The Universe Is Much Heavier than It Looks

Physicists have realized that something was amiss, the first time they seriously started weighing the universe. Obviously, they did not bring a huge scale along to do that. Rather, they inferred the distribution of matter inside and outside of our galaxy, by looking at the so-called gravitational lensing effect ($\nearrow$[1]): Images of far away galaxies appear distorted, because of all the other galaxies and black holes between us and them. These curve space and time ($\nearrow$[2]), and therefore bend the path of light rays. So the more distorted far away galaxies appear, the more mass there is in the universe, and the more matter there has to be. When the total matter content in the universe was estimated that way, it was way off what one would expect from the amount of stars and galaxies one could actually see. Even if one took into account all the dust and gas within and in between galaxies, a lot of mass was still missing: Space-time was clearly a lot more curved than expected.

Well, one could now guess that there had to be a lot of some kind of matter, which was quite massive and therefore curved space and time, but which was invisible to the naked eye, or to any other way of detecting electromagnetic radiation. Actually, it could not interact by any of the other forces ($\nearrow$[3]) except gravity, so it would also have to be completely pervious – one could not even touch it.

But could it not also be that one had simply made some mistake estimating the matter that should be there? Maybe the equations for gravitational lensing were not quite right? Theoretically, that could be possible. However, there are other, independently, indications that there must be more matter in the universe than one can see.

One such indication is that the outer regions of galaxies rotate much faster than they realistically should. Galaxies, no matter whether they are elliptic, irregular, or have spiral arms ($\nearrow$[4]), rotate like carousels. If the visible mass they contain was anything to go by, then the stars further away from their center should be much slower than the ones closer inwards. Galaxies should rotate a lot like, for example, our solar system: The outer planets like Saturn and Jupiter are much slower, and need a lot more time to orbit the sun than the planets further inwards, such as Mercury or Earth. But with nearly

$\nearrow$[1]: *"Gravitational Lensing"* on page 129
$\nearrow$[2]: *"Curved Space Time"* on page 125
$\nearrow$[3]: *"Standard Model of Elementary Particles"* on page 213
$\nearrow$[4]: *"Galaxy Types"* on page 113

all galaxies, that is not the case. Rather, the portions at the fringes move with the same speed as the core regions. It is as if there was a large amount of invisible mass that was dragging the outer spiral arms of the galaxies along with it, by pulling at them with gravitational attraction.

This list of indirect indications goes on: For instance, the way the whole visible matter is arranged in the universe is very hard to explain, without large amounts of invisible matter. Simply put, if there were just ordinary matter in the universe, it would not have had enough time since the Big Bang ($\nearrow$5) to arrange itself in the pattern we see it in today. Rather, the way it is today is far more clumped together ($\nearrow$6) than it should be. A large additional gravitational pull by some additional – invisible – matter can explain this structure.

The Hunt for Dark Matter: What Could It Be?

There have been several suggestions for explaining some of these phenomena. For instance, a slight modification of Newton's second law can explain the abnormal way galaxies rotate. But it cannot explain the other strange occurrences we have described

above. To the present day, the only explanation for all these phenomena is that there is a strange type of matter, which only interacts via the gravitational force, but not (or only very, very weakly) with the other forces. This hypothetical kind of matter has been termed dark matter, although whatever it is, it is not dark, but rather transparent.

Dark matter also seems to be able to pass through itself and other types of matter (which is just another way of saying that it does not interact via the electromagnetic force). The best example for this is the so-called Bullet Cluster: it is the image of two colliding galaxy clusters. From gravitational lensing effects with all the matter in the background, one can infer the mass distribution. The center of mass is at a completely different position where the center of the radiating matter is. It is as if the normal matter in the two galaxy clusters has been colliding, while the dark matter around it has just continued on its path, traveling undisturbed.

We can see it in the picture on the next page: Red indicates the visible mass of two colliding galaxy clusters. They form shock fronts (hence the reference to a "bullet") while passing through each other. The blue part indicates the dark matter, as inferred by gravitational lensing. It passes just through the visible matter and each other undisturbed.

So what exactly is this dark matter? For quite some time, physicists have had the theory that it could have been neutrinos ($\nearrow$7). For a long time, people did not know whether neutrinos had any mass or were massless. If they had had a not too small mass, it could have been that dark matter was in

$\nearrow$5: "The Big Bang" on page 97
$\nearrow$6: "Large Scale Structure of the Universe" on page 109
$\nearrow$7: "The Neutrino" on page 209

fact just the large amount of neutrinos present in the universe. Neutrinos are quite weakly interacting, so that could have been a possibility. Alas, since the neutrino oscillation experiments (↗8), it is known

point of writing this article, it still has not been confirmed experimentally. Rather, the more experiments the Large Hadron Collider (↗10) performs, the more and more unlikely it appears.

that neutrinos have indeed a mass – but that it is so incredibly tiny that all the possible neutrinos in the universe cannot explain the large amount of excess matter. So neutrinos are out as a dark matter candidate.

Another possibility which has been entertained for quite some time now, is that dark matter particles might actually be the lightest supersymmetric particles (↗9). Supersymmetry as a theoretical concept is very appealing and explains many gaps in the standard model of particle physics. And, possibly, also gives a dark matter candidate. But as of the

So up to the present day, the precise nature of dark matter is still a mystery. Is it some strange supersymmetric particle type? Does it belong to some completely unknown, yet to be discovered species of elementary particles? Is it maybe just an error in our equations for gravity, leading us to believe that something must be there when it actually isn't? Is it maybe a strange quantum gravity effect (↗11) we haven't anticipated?

We only know that something is amiss. And until we find out what exactly it is, this will remain one of the unsolved puzzles of our universe.

↗8: *"Neutrino Oscillations" on page 261*
↗9: *"Supersymmetry" on page 299*
↗10: *"Particle Accelerators" on page 249*
↗11: *"Quantum Gravity" on page 279*

Image (composite): X-ray: NASA/CXC/CfA/M.Markevitch et al.; Lensing Map: NASA/STScI; ESO WFI; Magellan/U.Arizona/D.Clowe et al.; Optical: NASA/STScI; Magellan/U.Arizona/D.Clowe et al.

DARK ENERGY

Dark Energy
The Revival of Einstein's Biggest Blunder

We know that the universe is expanding since Edwin Hubble measured the distance and redshift of far away galaxies (↗[1]). For many years, many people thought that the question of whether the universe would expand forever or not, would come down to how much matter and energy is contained in it. There are three cases: the "closed", the "open" or the "flat" case. In the closed case, the universe is so heavy that it will eventually collapse back to a point under its own weight – very much like the Big Bang in reverse; in the open case it is so light that it will expand forever, but more and more slowly, growing beyond all boundaries; the flat case is something like the marginal case between the two: the total mass of the universe is just right so that the universe will grow indefinitely, but so slowly that it will not grow beyond a certain point. The names "closed", "open" and "flat" come from the geometry the universe will have in each case: It would be curved positively, not at all, or in a negative way, respectively.

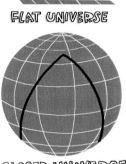

FLAT UNIVERSE

CLOSED UNIVERSE

OPEN UNIVERSE

Standard Candles in the Dark

Physicists presented these three as the possible cases for the universe, because these are the three possible solutions that come out of the equations of general relativity. Well, that is, when you assume that there are just space, time and matter (including the dark one ↗[2]) in the universe.

But this view of the fate of the universe changed, when in 1998 observations were published, which were concerned with the signals from exploding supernovae, specifically those of type Ia (↗[3]).

These specific explosions of stars are fantastic for astronomers! (Those on Earth, that is. Probably not for any astronomers that happen to live on planets nearby a type Ia supernova.) They happen all over the universe, throughout most of its history. Because the flash from the explosions travels at the speed of light in an expanding universe, we can actually tell very well how much the space has expanded since the event happened. Also, the type Ia supernovae always flash at a very specific brightness – so by comparing it with the actual amount of light that reaches us, we can guess how far away the supernova is about now.

These data were collected over several years, and for numerous supernova explosions – the further they were away, the longer they had happened in the past, painting a detailed picture of the evolution of the cosmos over the past couple of billion years.

Doing this, the astronomers found out something very astounding about our universe: Not only is the universe expanding now, it is expanding faster than it has been in previous times. Unlike any of the three

↗[1]: *"The Big Bang" on page 97*
↗[2]: *"Dark Matter" on page 133*
↗[3]: *"Supernovae" on page 83*

possibilities that physicists had believed to be possible till then, the expansion of the universe is actually increasing!

And what about the shape of the universe? Is it closed, open or flat? Well, also that question could be answered with a certain precision at the end of the last century: By careful observations of the way the ripples in the cosmic microwave background are distributed (↗4), physicists could demonstrate that the universe is, to a very large degree of accuracy, flat.

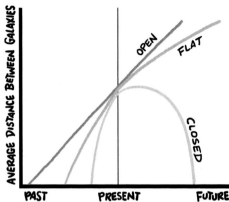

The universe Is Flat – and Expands Faster and Faster!

This does not, of course, at all fit with the original three possibilities, in which a flat universe meant that the universe had to expand slower and slower, to eventually reach a maximum size. But there is a bigger problem here: In order for the universe to be flat, it needs to be heavy enough (more precisely: to have a specific, called critical, density). But if we look around us and count stars, galaxies, nebulae, dust, photons, etc., we don't get nearly enough matter and radiation to reach the critical density – only about 4.9% of what is necessary to keep the universe flat. With this amount of stuff in it, it should totally be an open universe. But it is clearly not, as the CMB measurements show!

This may sound to you like something you might have read before: isn't there lots of matter we cannot see, but we know should be there from other considerations? Yes, the so-called dark matter (↗2) – could that make up for our missing density of the universe? Well, as it turns out, there is more dark matter than ordinary matter in the universe. But dark matter and ordinary matter together only make up about 31.7% of the critical density in the universe. Also, whatever this additional density is, it needs to be extremely homogeneously spread out over the whole universe – otherwise we would notice it in gravitational lensing. So unlike dark matter, which tends to clump around ordinary matter, this missing whatever would need to be everywhere, equally finely distributed. Even in the vast emptiness between galaxies, where almost nothing can be found.

Actually, in Einstein's equations of general relativity (↗5,6), energy also curves space and time. If the universe was filled homogeneously with a certain type of energy, then that would bring enough energy density to make it flat, rather than negatively curved. This missing energy is also called dark energy, in analogy to dark matter – and remains even more mysterious than the former.

As If Dark Matter Weren't Enough We Didn't Understand!

There is one excellent way to explain dark energy, which does not only answer the question of why the

↗4: "The Cosmic Microwave Background" on page 105
↗5: "The Theory of General Relativity" on page 121
↗6: "E=mc²" on page 237

universe is flat, but also why its expansion is accelerating. You just have to add an additional term to Einstein's equations, called the "cosmological constant". This cosmological constant behaves like an energy, but it drives the universe to expand. Einstein realized that a static universe would collapse due to his original set of equations. And since he wanted the universe to be eternal, he originally added the cosmological constant to counteract the expansion. It was just later, when he found out that the universe was, in fact, not static, but expanding, that he abolished the additional term, calling it his "biggest blunder".

But guess what: a term with a cosmological constant is precisely what one would need to add to the equations in order to explain dark energy: it behaves like an energy, so it curves space and time the right way. But it also has negative pressure, so it drives the universe to expand more and more rapidly! Because this is a constant term, it is the same everywhere – which would also explain why it is completely homogeneously distributed over the whole visible universe. Actually, the model which, at the time of writing this book, agrees with all cosmological measurements within the observational errors is the so-called ΛCDM (pronounced "Lambda-Cee-Dee-Emm") model. It got its name, because it includes a cosmological constant, the symbol for which is the Greek letter Λ, as well as "cold dark matter". It is also called the "standard

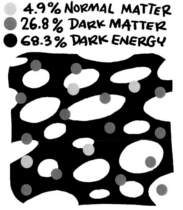

4.9% NORMAL MATTER
26.8% DARK MATTER
68.3% DARK ENERGY

model of cosmology", because it fits nearly perfectly with our current observations.

Einstein's Biggest Blunder – Revived as Dark Energy

That could be the end of the story – the dark energy is simply a cosmological constant term in the equations. But there are some aesthetic concerns with this explanation. The more speculative branches of physics, such as String Theory (↗[7]), usually predict a cosmological constant which has a value of around 1 (in natural units). The value that one would have to use in the ΛCDM model to agree with observations, though, is 10^{-120}. That is a zero, a decimal, 119 more zeros, and then a one! That is a number which is so astronomically tiny that it seems ridiculous to some physicists. Why should any constant of nature have such a small value? In fact, there have been some other attempts at an explanation, with exotic kinds of matter fields that fill the universe. Technically, these would behave very similar to a small cosmological constant. But some people would feel much better, knowing there was an additional type of (yet unknown) matter in the universe, explaining why the universe is flat and expanding faster and faster.

At the moment, however, the only thing we know for sure is that a cosmological constant explains a lot of the how in cosmology, but not really much of the why. Whether that is a sensible question to ask, or one for which we will ever have a satisfactory answer from physics, is a completely different matter.

↗[7]: "String Theory" on page 303

III – Quantum Mechanics

"We live in a special world – everything we are used to and deal with on a daily basis, is large. Chairs, houses, cars, even we ourselves, are huge, compared to the atoms and molecules we all consist of. In fact, we are so large, that we cannot see, hear, or experience in any other way, a single atom, at least not without technological help. That is why we were in for a surprise about a hundred years ago, when technology had advanced enough so we could begin to access the tiny world of the atom. For the first time, we could manipulate the building blocks everything is made and constructed out of – and they were weird. Up to that point, we had imagined atoms to behave just as the other macroscopic objects we were used to, like chairs, or houses. Essentially, we had imagined atoms like little rocks, or marbles.

But they weren't! It turned out that atoms and other particles were ruled by completely different laws than the objects we experience in our daily life. Even of a deeper, philosophical level, the microscopic world operated on a completely different level than our macroscopic world. We just had not noticed this, because we are so freakishly huge in comparison to a single atom.

This microscopic world is ruled by the laws of quantum mechanics, and its strangeness has befuddled and inspired scientists and laypeople for over a hundred years. Whether an atom can take two different paths to its goal at the same time, have a spooky interaction at a distance with another particle, or overcome obstacles by simply tunneling through them, the strangeness of the quantum world never ceases to astonish us. On the next couple of pages, Maxwell, Emmy, and myself will try to tell you about the most astounding and counterintuitive phenomena that can be found in the realm of quantum mechanics."

Wave-Particle Duality
Is It a Wave or Particle?

Every piece of matter in our world consists of molecules. These molecules are collections of several different kinds of atoms sticking together. With enough energy one can break atoms apart into their constituents, the elementary particles (↗1). As far as we know today, the elementary particles are themselves not composed of anything, but form the fundamental building blocks of matter.

What are these elementary particles like? Are they "things"? Our intuition for "things" is very particular – and we are so used to it that it is hard to imagine anything different.

Macroscopic objects appear to have a number of important properties: They are "solid", so where one thing is, another cannot be. This is closely connected to the notion that a thing is always somewhere, and nowhere else. All right, an object can be extended and have a certain volume, but then we can actually break it into several pieces. So rather, we can think of it as several objects sticking together. Something similar is true for liquids: one can pour one into the other, but we secretly know that this is just mixing the little particles that make up the liquid.

Another important feature is that "things" have certain physical properties, such as position (i. e. where it is), velocity (how fast it moves), and several others like energy, angular momentum and such. Even if we don't know the value of one of the properties, the "thing" always has these properties.

Electrons Behave like Waves …

Unfortunately, nothing of what we have described above is true for elementary particles! Take an electron, for instance. It is one of the lighter particles, and the carrier of electricity. If an electric current flows in a cable, it is because the electrons in the wire are moving in one direction.

An electron – as well as any other type of elementary particle, too – is a quite curious thing. Rather than behaving classically, it obeys the laws of quantum physics. For instance, it does not have a certain position. That means that it isn't just "here but not there", but there are only certain probabilities, with which to find the particle at some point. These probabilities are collectively described by the wave function. Think of this as a water wave that travels on the sur-

↗1: "Atoms vs. Elementary Particles" on page 205

face of a lake. At those places where the amplitude of that wave is very large – in other words, where the water surface is moving a lot – the probability of finding the electron is very high. At those points where the amplitude vanishes – where the water surface is perfectly still – the electron is most likely not to be found.

The analogy with a wave goes even further: The probability of where to find the electron can be scattered, reflected, made to interfere, and many more things that one can also do with water waves. A notorious example for this strange behavior is described in the article on the double slit experiment (↗2).

... but Also like Particles!

Mind you, this does not mean that one can split up the electron into two halves, by splitting up its wave function. There is only ever one electron, and it always has the same mass and the same charge. One actually can split up the wave function of the electron. This means, one can make part of the wave travel in one, and some other part of the wave travel in another direction. But that just means that, when we take a look to actually find out where the electron is, we'll find the whole electron either in one or the other direction, with a 50-50 chance for either possibility, if we have split the wave function evenly.

Which One Are They? Both!

The wave-particle duality of electrons (and other elementary particles) can be summed up the following, very simplified way: The electron is a single, indivisible particle. But the way that particle moves from one point to another is very similar to how waves move.

There are actually many more strange properties of quantum particles, which are in one way or another related to this wave-particle duality. For instance, one can make an electron go two different ways at once (↗2), and it can spontaneously cross barriers it normally wouldn't be allowed to (↗3). It is very difficult to pinpoint exactly where the electron is, at any time, at least if you also want to know how fast it is (↗4), and quite generally, an electron can be at several different places at the same time – at least until you look close enough (↗5).

So why is, say, a chair not weird, if it consists of weird particles?

SKATEBOARDING PROHIBITED

↗2: *"The Double Slit Experiment"* on page 147
↗3: *"Quantum Tunneling"* on page 163
↗4: *"Heisenberg Uncertainty"* on page 151
↗5: *"Schrödinger's Cat"* on page 155

All right, we have now established that the microscopic world is different from the macroscopic world: the objects of our everyday life behave classically, but elementary particles follow the laws of quantum physics instead. But wait – didn't we say in the beginning that macroscopic objects such as doors, chairs, rocks and brownies, are, fundamentally, composed of lots and lots of elementary particles? So why is it that something that consists of quantum particles, does not behave in a quantum way? Why does a rock not behave like an electron?

To answer this question, it is good to know that the usual spread of the electron wave function in e. g. a piece of rock is roughly a few Ångströms, i. e. 10^{-10} m. Remember, by the way, that this is not the size of the electron, but the size of the region where we can most likely find the electron. The wave functions of the other particles in the rock, such as protons and neutrons, are even smaller by a factor of about 2000. But the rock itself consists of millions of billions of billions of particles (10^{24}, an enormous number), and is usually a few centimeters in diameter.

So the rock itself has relatively little quantum fuzziness: The particles in it jump back and forth over the distance of a few Ångströms, but on average the rock stays where it is. Maybe there is a little quantum uncertainty of where its surface is (i. e. where the rock ends and the air begins), but that uncertainty is again of the order of magnitude of a few Ångströms, at most. That is far too small for us to notice.

Another example: If you hit a not too thick wall with an electron, there is a certain probability that it vanishes and appears on the other side (called tunneling, ↗[2]). Say this probability is 10%. If you throw a rock at a wall, then, for the rock to vanish and appear on the other side, all electrons have to tunnel through the wall! The probability for that to happen is 10^{-24}, or 10^{-22} %! So if you have been throwing stones at a wall literally since the beginning of time (14 billion years, i. e. $4 \cdot 10^{17}$ seconds), the chance for a stone to have tunneled through the wall at some time is still roughly one in ten million! So it is just extremely unlikely to observe macroscopic objects behave in a quantum way.

The Double Slit Experiment
On the Weirdness of the Quantum World

Quantum mechanics is weird. As Richard Feynman famously said: "Nobody understands it." With this he did not mean, by the way, that we can not make precise predictions with the laws of quantum physics – we can – he just meant that the laws themselves are pretty strange. In fact, they contradict the intuitive understanding of the world, that we as human beings intrinsically have. Sounds pretentious, but unfortunately that's how it is. The prime example of the strange behavior of quantum particles is demonstrated in the so-called double slit experiment.

The double slit experiment is famous among physicists: It is one of the most direct evidences that microscopic elementary particles behave very differently from the macroscopic objects of our everyday life experience. To the present day, most introductory courses to quantum mechanics begin with the description of this experiment. By the way, it is not just a thought experiment like the one involving Schrödinger's cat. The double slit experiment has been performed many times in the past century, in order to demonstrate the weirdness of quantum physics.

Experimental Setup: Don't Let Them Electrons Get Through!

The experiment goes like this: Imagine a wall, with two vertical slits next to each other, through which you can peek to the other side. In front of the wall, you mount an electron cannon, aimed at the two slits. If you fire up the cannon, it will emit a steady stream of electrons, flying towards the wall. Behind the wall, you put up a dark, electrosensitive screen. Every time an electron hits it, a little white dot appears, so you can see exactly where the screen has been hit.

The question is: if you switch on the cannon in front of the wall, what pattern will you see on the screen behind the wall?

Okay, let's make the question easier for a start: Cover one slit up, for instance with tape (anything that doesn't let electrons through will do). So you effectively have a wall with one very slim hole in it. You fire up the electron cannon, then what happens? Well, most of the electrons will crash into the wall. But some will exactly hit the slit, and travel through the wall. These will hit the screen on the other side, and produce a tiny bright spot on it. If you leave the cannon on for long enough, the pattern you will see on the screen will be just a copy of the slit (in a sense its inverted shadow). Also, if you open up the covered slit, close the other one, and repeat the experiment, you will see a second copy of the slit appearing, next to the first one. So far, so predictable.

All right, now to the strange stuff: What happens, if you open up both slits at the same time? The natural answer every normal person (including all physicists before the end of the nineteenth century) would give to that is: Well, those electrons which do not crash into the wall, will go either through one or the other slit. So with both slits open, the screen will show the images of both slits next to each other.

Wrong! That is not what you see. Rather, a so-called interference pattern emerges on the screen. It looks like a bit like a zebra crossing, going from left to right: There are several vertical lines, where lots of electrons have hit, next to each other. Between every two neighboring lines there is a region where no electrons have hit, where the screen stays dark.

Interference: A Particle Interacts with Itself from Another Reality

The reason why this strange pattern emerges, lies in the quantum properties of the electrons. Although it is very appealing to think of particles as little dot-like clumps of matter that fly around in straight lines, that is not at all how they behave. Rather, the probability of where an electron exactly is, behaves like a wave. It is described by the so-called wave function (↗1).

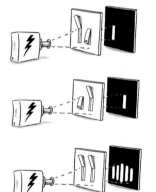

In particular, it is extended in space: the electrons are not just at one specific point in space, but they can be found at different places with certain probabilities. So the wave-function of an electron can go through both slits at the same time. Moreover, the two parts of the wave function that emerge behind the wall interfere with each other – as befits waves, which can do this sort of thing – causing the interference pattern on the screen behind it.

If the two parts of the wave function hit the screen, two extreme cases can happen: Either, both waves are at their maximum (or both at their minimum, that's the same thing here), and the resulting wave function, which is the sum of the two, has a really large amplitude. At those places, the electron is very likely to hit the screen, and this is called constructive interference. The other possibility is that one wave is at its maximum, and the other is at its minimum. They cancel out, and the amplitude of the wave function is zero. You guessed it, that is called destructive interference. At these places, the probability of the electron hitting the screen is zero. Those regions on the screen are a bit like spots in a room with bad acoustics: sound waves interfere destructively with each other, and as a result one can hear next to nothing when standing there. You have all found that precise spot in a concert hall.

Wait a minute, that's nonsense – I hear you say – an electron cannot go through both slits at the same time. The electrons have to go either though the left slit, or through the right slit! And then, behind the wall, those that went left, bump into the ones that went right in some way, and that somehow creates that strange pattern when both slits are open at the same time. Couldn't that be an equally valid interpretation of the result?

An intriguing thought, but it's not what happens. Here is why: You can tune down the output of the electron cannon to such a low intensity, that it emits only one electron every couple of seconds. So there is never more than one electron in the air at any time. Again, many of these electrons will hit the wall, but sometimes one will

↗1: *"Wave-Particle Duality" on page 143*

hit the two slits and produce a single dot on the electrosensitive screen. After a few hours, when lots and lots of dots have appeared, one can again see the interference pattern emerge on the screen. So this pattern cannot be a result of two or more electrons bumping into each other – there is only ever one electron flying. And this one electron goes through both slits at the same time, interfering with itself in the process.

In fact, you can shoot just one electron at the screen in your laboratory, and do this experiment in several other laboratories all over the world as well. If you place the images of the screens from all labs (every one with at most one dot) on top of each other, you will see the same interference pattern as a result.

Don't Peep: If You Look, There Is No Interference

All right – you might say – but I still don't quite believe it. What if one installs a little detector in the slits, which checks through which one the electron flies? Ingenious engineers that we are, we can make these detectors so sensitive that they just register which slit the electron passes through, without disturbing its path.

Yes, one can build these detectors, and one can repeat the experiment with it, noting for each electron which of the two slits it passes trough. If we do this, we can see that the electrons go through either of them with a 50-50 chance. Sometimes the electrons go through one, and sometimes they go through the other. But we also note something else: The interference pattern on our screen vanishes! Instead, the pattern that emerges is the one

we would have expected from the beginning, had we not known anything about quantum mechanics: The image of the two slits, next to each other.

Here's the deal with quantum mechanics: Because we received information about which of the two paths the electron was taking, it stopped behaving like a wave for a moment, but behaved like a particle. For an instant, it actually acted like the small cannon ball, which fits into our world view of "particle", and went through either of the two slits, rather than through both at the same time. And because of that, the pattern on the screen actually looks like the electrons were particles in the usual, classical sense, without wave properties.

But it only did that because we measured its particle properties, i. e. its position, at the slits. If we remove the detector at the slits in the wall, the interference pattern returns. This is a prime example of how the measurement process actually disturbs the system it measures, and this is one of the key features that is often associated with quantum mechanics.

As hard as all of this is to believe, it is actually true. To repeat it once again: these experiments have been performed time and again, and the predictions of quantum mechanics have always been confirmed, precisely, without exception. So even if the laws of quantum physics contradict many of the intuitions we have as human beings, we need to come to terms with the idea that this is just the way of nature: fundamentally it is more complicated and wondrous than we imagined. The universe just doesn't do us the favor of conforming to our expectations.

"Location is overrated. How about this beauty of uncertain address that's probably close to something, or other?"

Heisenberg Uncertainty
You Cannot Have It Both Ways

We have mentioned it already, and probably will a couple more times again in this book: the quantum world is weird. The elementary particles that constitute all matter (that we know of, at least), behave in a very weird and counterintuitive way. At the heart of all this lies the wave-particle-duality ($\nearrow$[1]). Although the elementary particles are indivisible and fundamental (again, as far as we know), they do not move about like tiny, hard balls. Instead, the way they move and interact is far more like waves.

Position and Velocity: You Cannot Know Both Precisely

One of the more well-known consequences of this is the famous Heisenberg Uncertainty Principle. It is often stated in the following way: "One cannot measure the position and the velocity of a particle at the same time." Another way is to say that "Every measurement process changes the object one measures. If you want to measure where the particle is, you inevitably influence it in such a way that you don't know how fast it is anymore."

This sounds very intuitive, which is why this explanation has been brought up again and again throughout the decades. But it misses the point. It is true that every measurement of a quantum particle influences the state of that particle, but that has nothing to do with the Heisenberg Uncertainty principle.

The Heisenberg Uncertainty Principle states that a particle does not have a well-defined position and velocity at the same time. This is not a question of how much we do or don't know about it, but about what the actual state of the particle is.

It's All about the Waves, Baby

At the core of this lies the fact that particles behave like waves. More specifically, the probability of finding the particle somewhere if one looks for it, is largest at those places where the amplitude of that wave is large. The velocity of a particle, however, is large if the frequency of the wave is large (equivalently, if the wavelength is small). If we know this, then we can explain why a particle cannot have, at the same time, an exact position and an exact velocity.

Frequency and Amplitude

Assume that it were very certain to find the particle in a certain, small region. In other words, in that region the wave function would have a very large amplitude. Outside of that region the wave function would have to be very small. It would look like a sharp, narrow spike, which was concentrated at a very specific point. That would be that point where we could find the particle, if we were to look for it.

Now, it is a fundamental property of waves, that any single wave usually contains many frequen-

$\nearrow$[1]: "Wave-Particle Duality" on page 143

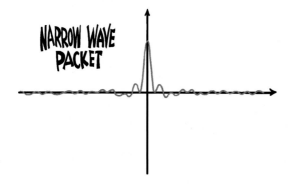

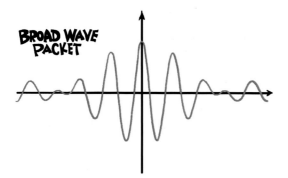

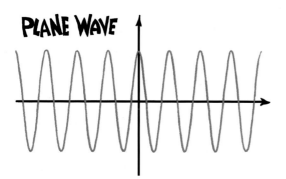

cies. Just as white light coming from the sun contains all the colors of the spectrum (and therefore, electromagnetic waves of all sorts of wave length, ↗²), wave functions contain many different frequencies. And it is a fact of the way that waves work that, in order to concentrate a wave to a very small region ("narrow wave packet"), it needs to contain lots and lots of different frequencies. You need ever more frequencies the more narrow you want to confine the wave to a certain region.

This is the core of the Uncertainty Principle: a particle that can be found, with a very high probability, in a very small region, needs to have a wave function which contains a huge number of different frequencies – if the position is very certain, the velocity is very uncertain. This also works the other way round: Take only a few different frequencies (and hence just a few different velocities) and you will get a localization across a broader space ("broad wave packet"). If you want to have a particle which has a very certain velocity, you need to take a wave with only one frequency ("plane wave"). But that is a sine wave, which is extended infinitely in all directions, having the same amplitude everywhere. So if you want to know the velocity of a particle for certain, its position has to be completely unknown.

The Musician's Uncertainty Principle

What we have just described, by the way, is not just a weirdness of the quantum world – it is true for all waves, even the ones we experience in everyday life!

A good example for this are sound waves. There is, in fact, a "Musician's Uncertainty Principle". With sound waves, the amplitude tells you how loud the

↗²: "Light" on page 7

wave is, and the frequency tells you about the pitch of the note. So, here it goes: "If you want to hear which note someone is playing precisely, you have to be able to hear the note for a certain period of time. That time is longer, the lower the played note is."

The human ear is quite good at distinguishing frequencies of sound waves, but, because of the physical nature of the waves, it needs a certain time to do this.

There is a different way to say this: If you want to be able to distinguish two frequencies which are very close to each other, you need to listen to the sound longer. The more precisely the frequency is known, the less concentrated the wave is allowed to be (i.e. the longer you have to play the note).

Actually, this is quite important for the people playing in an orchestra, and the musicians there all know this, even if only from their own intuition. Those instruments which play only very short notes don't need to be tuned very precisely. Simply because a short time does not allow the human ear to hear, whether the precise frequency has been played or not. On the other hand, those instruments which usually have to play long notes (like violins, for example), need to be tuned precisely. The longer a note is played, the more easily can one hear when several violins do not all play exactly the same one.

You can test the musician's uncertainty principle for yourself online, if you like (➚³).

Quantum Weirdness: Even Without Measuring Anything

So the Uncertainty Principle is actually not very uncommon – and it has been known long before Heisenberg. Well, it still came as a surprise to the physicists in the beginning of the 20th century. Not because of the fact that waves have this uncertainty principle, but rather, that particles behaved like waves, and the uncertainty principle for particles meant that particles cannot have an exact position and velocity at the same time.

Next to the formulation of the Uncertainty Principle that uses the uncertainty on a particle's localization and velocity, there are also others. You can also define it as an uncertainty on a particle's energy and the time it has that particular energy (similar to the musician's Uncertainty Principle).

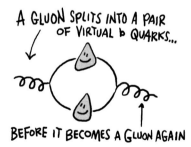

A GLUON SPLITS INTO A PAIR OF VIRTUAL b QUARKS...

BEFORE IT BECOMES A GLUON AGAIN

The smaller the time interval, the more deviation from a certain energy a particle can have. Take a gluon, for example. For a very short amount of time, it can "borrow" some extra energy and split into a pair of quarks. But after some time, these quarks have to convert back to a gluon and give their energy back. That's why such particles, that only exist for a certain amount of time, are called "virtual particles". The shorter the time interval, the more you can do and hence the more particles you can produce virtually.

➚³: http://newt.phys.unsw.edu.au/jw/uncertainty.html

"What do you mean he's not really dead until we open the casket?"

Schrödinger's Cat
Dead and Alive at the Same Time

We have talked a lot about the weirdness of the quantum world (↗[1,2,3]). A lot of that weirdness comes from the wave-particle duality: On many occasions, elementary particles show wave-like behavior.

But that is not the only reason why the quantum world is so strange to us. In fact, the wave-like nature of particles is something that we have understood quite well by now. But there is another part of the quantum world, which is actually very counter-intuitive, and outright mysterious. That is the fact that we cannot see that wave-like nature directly – we can only observe its consequences.

Superpositions of States

You might have already read about this in the article about the double-slit experiment (↗[2]). A quantum particle can be in a superposition of states. A physical property that it has can have several different values simultaneously! For instance, a particle can take two different paths at the same time. That in itself is not surprising for a wave – but what is so very mysterious is that we invariably destroy this superposition, when we actually make a measurement, e.g. to check where the particle is. We can never see the superposition directly: whenever we look closely at the particle, it "decides" for either one or the other possibility. And there is no way to determine in advance which it will take – we can only compute the probabilities with which each will be chosen.

Schrödinger and His Cat: a Gedankenexperiment

Wait a minute! – an Austrian physicist from Vienna named Erwin Schrödinger, said in 1935 – this is actually really strange. He devised the following gedankenexperiment ("thought experiment" in German): If you have a radioactive material (↗[4]), that has a certain half-life. This is a certain time, after which (on average) half of the material has decayed. What happens if you only have a single atom of that material? Well, in the beginning, it will be in the state "particle is there". After the half-time, however, it will be in the quantum state which is the superposition of "particle is there" and "particle has decayed". It will be both there and not there, in a quantum superposition of the two possibilities.

But now, Schrödinger asked, what happens when we put this radioactive atom in a box, together with a detector, which does the following: when it detects radiation from a decay, it releases a deadly poison into the box. Also,

↗[1]: "Wave-Particle Duality" on page 143
↗[2]: "The Double Slit Experiment" on page 147
↗[3]: "Heisenberg Uncertainty" on page 151

↗[4]: "Radioactive Decay" on page 167

put a cat into the box. Then close the box and make sure that it is tightly sealed, so that no light or sound can come out of it.

Now, Schrödinger argued, after the half-life has passed, the radio-active nucleus is in a superposition of having decayed and having not decayed. So the detector would have to be in a superposition of having released the poison, and having not released the poison. The cat itself, therefore, would have to be in a superposition of both being dead, and being alive.

So what happens when we open the box? The cat is both dead and alive, and as soon we look inside, the universe "decides", and we see either the dead cat, or the alive cat? That would be totally strange!

The Cat Is Dead and Alive at the Same Time – but Why?

Back in his days, Schrödinger presented this idea, to show how very strange it would be to apply the ideas of quantum mechanics to macroscopic objects of the real world. But it highlights another important question of quantum physics, which we do not really understand: where is the divide between the quantum and the classical world? Or, differently put: what does it mean to "measure" something?

One could, for instance, say the following: Schrödinger's cat is not in a superposition of quantum states, because of the detector. The detector measures whether the nucleus has decayed or not, so that is where a measurement happens. The poison is either released or not, but not both.

Well, that is a compelling argument. But unfortunately it is wrong. That is indeed something one can show with an experiment, which is called the "quantum eraser". This is an experiment (fortunately without potentially murdering cats), in which a quantum superposition of two states is produced. Then, a measurement is performed, to check which of the two states is actually realized. By the rules of quantum mechanics, that should destroy the superposition. But what one actually finds is the following: If one deletes the information of the measurement result, without ever looking at it, then the superposition is not destroyed. So it seems not to be about whether a measurement is performed. It seems to be about whether anybody knows about the result of the measurement!

But that is really strange. There is no good physical definition of "knowing". One could rephrase the question of Schrödinger's cat the following way: Why does a measurement take place when the scientist opens the box, but not when the cat realizes that the poison is released? A cat can know something, right? So can the cat perform a measurement?

Wigner's Friend: Is He Also in a Superposition?

Actually, if one takes the notion of "making a measurement" literally, one can take the gedanken-experiment of Schrödinger's cat to the next level. The idea for this comes from the physicist Eugene Wigner. It goes like this: Wigner wants to perform the experiment with Schrödinger's cat, and he has set up everything for it. But Wigner is a squeamish guy, and is afraid of finding a poisoned cat. So he waits in front of the lab, and sends in his friend. That friend then looks inside the box, sees whether the cat is dead or alive, and then leaves the lab to tell Wigner the result.

What happens in the few seconds between Wigner's friend opening the box, and him going outside, to tell Wigner? If one takes the notion seriously that the measurement process is not about a mechanical process, but about whether a certain information exists somewhere or not, then one has to conclude the following: The moment Wigner's friend opens the box and looks inside, the combined system "cat and Wigner's friend" is in the superposition of "cat is dead and Wigner's friend decides to tell Wigner to stay outside" and "cat is alive and Wigner's friend can tell Wigner it is safe to come inside and see for himself". It is at the moment that the lab door opens that the wave function of the combined system "cat and Wigner's friend" collapses, and the universe decides which of the two possibilities is realized.

Actually.... where does that stop? Has Wigner already told you about

the result? No? Well, then until you go and ask him, the physical system of cat, his friend, himself, and all the people he has already told, and the people those people told, and so forth, should be in a huge quantum superposition of "a lot of stuff that happens after the cat dies" and "a lot of stuff that happens after the cat survived". And as soon as you find out, you become part of this huge quantum superposition.

Phew! At this point, the whole debate becomes quite metaphysical, which is why most physicists give up, and go have a drink or three, until the headache goes away.

Nowadays, still nobody has actually understood what "making a measurement" really means, in the context of quantum mechanics. Where precisely is the boundary between the quantum and the classical world? Between microscopic and macroscopic? Maybe we will never know. But it is safe to say that Schrödinger's cat has inspired and mystified generations of physicists alike. It serves as a reminder that we haven't really understood quantum physics, in all its strangeness.

"So what you're saying, Professor, is that this isn't your actual lecture on path integrals. This is just the <u>directions</u> to your lecture on path integrals."

Feynman Paths
Reality as the Sum of Possibilities

Imagine you are walking along the beach. It is a beautiful day, maybe even a little too warm. Suddenly you see an ice cream cart. Excellent! But wait: it seems as the ice-man is scratching the last bits of ice cream out of his boxes. No time to lose! Which way would you go if you wanted to get to the cart as fast as possible? The direct way, for sure.

If there is just sand on the ground, this direct way is also the fastest. But the situation can change if part of way is covered with asphalt. There, you can run faster, but the way would not be that direct. Intuitively, you would optimize your way and run part of it on the asphalt, part on the sand. This would no longer be the shortest route, but the fastest. Did you know that light follows the same principle? Looking at the path of light will show you that it always takes the fastest way. Following this principle can explain some fundamental laws of optics. You could generalize the statement to "Hamilton's principle" that states that the action for each process is either minimal or maximal. That statement is quite abstract but means, roughly speaking, that everything that happens in nature is optimized.

But let us go back to light which takes the fastest way when traveling from A to B. This is called "Fermat's principle", named after the mathematician Pierre de Fermat who lived in the first half of the 17th century. The first example that we test is light, emitted from a lamp at point A and observed at point B. Between the two, we put a wall.

And below the wall, we place a mirror. Let us check all possible ways that light could go. The fact is, there is an infinite number of ways, but if we only take straight lines to the mirror and back to the observer, we get what appears in the illustration. Which one is the fastest? Right, the

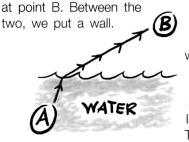

one that takes the way via the point below the wall. How do you know? Well, either you followed Fermat's principle and chose that path as it was the shortest, or you remembered the optical reflection law. It states that the incoming and outgoing angle of light reflected on a mirror are the same.

The second example corresponds to the ice cream cart and the asphalt. An observer at point B watches light that is emitted from a lamp placed underwater at point A. Given the fact that light is slower in water than in vacuum, it will no longer take the direct way. Instead, the fastest way is a compromise between a short way through the water and a fast way through the air. The light's path will look as in our illustration: it will have a kink. We know this effect as "refraction" of light. And the refractive index n, which you might know, is nothing but the velocity of light in a vacuum divided by the velocity of light in that medium.

From the Classical to the Quantum World

So far the answer to the question "Which way would light take?" is "The fastest!". Let us now change from the classical world to the world of quantum mechanics. One of the most important contributor in this business was Richard Feynman. The Feynman diagrams (↗1), which describe the interaction of elementary particles, were named after him. But more important is that Feynman did not just write

down little sketches, he also built, together with other physicists, the mathematical framework that is behind those sketches. For the theory of light and electromagnetic interactions it is called "Quantum electrodynamics" (QED). One way of performing calculations in QED, which means making mathematical expressions out of the nice diagrams, is via the so-called "path integrals". This formalism, quite complicated even for advanced physics students, will be introduced what follows. It might all sound very abstract and far from any daily-life, but it has really funny consequences and can explain quite a lot in a completely different way. We will follow the very illustrative explanation that Feynman himself used in his book "QED – The Strange Theory of Light and Matter".

Let us go back to the light that is supposed to go from A to B. Which way would it take? The fastest, you say? Feynman would have answered: "All of them!" Wait – how can that be? Indeed, using the path integral formalism you pick all the options you have, no matter how stupid they are. Take the mathematical expressions corresponding to all your options and add them. The result is something that we call "probability amplitude". As our quantum particles propagate as waves (↗2), they can also interfere. This means adding two waves does not necessarily result is a wave with a bigger amplitude. Instead, the waves can also cancel.

↗1: *"Feynman Diagrams" on page 225*
↗2: *"Wave-Particle Duality" on page 143*

We can illustrate that by assigning an arrow to each possible path that an object can take. The arrow moves like the hand of a clock, rotating with a speed that gets higher the more energy our object has. If we now want to add all possible paths of that particle, we have to add all the corresponding arrows. You can add them by putting the end of one to the point where another one starts. Depending on the direction into which they point, two arrows of the same length can add to a total arrow which is twice as long or they cancel completely. What is important: the probability amplitude – the total probability of a process to happen, corresponds to the length of the added vectors. So the longer the resulting arrow, the more likely a process is to happen.

If we take a second look at our mirror – now from the Feynman path point-of-view – we see that the arrows for each path point into different directions. If we add them all, we see that many cancel, in particular those which are far away from the center. You could say that "summing a bunch of random arrows" leads on average to zero. But around the center, the different ways do not differ much concerning their length. That's why all arrows there point into more or less the same direction. And this part is what contributes to the reflection, the part that "gives the total arrow its length". So by using a com-

pletely different approach we could also find out that the light will prefer the ways with the shortest distance.

You can also use the arrow technique to explain why light is taking a straight path when staying in one medium. Close to the straight line, the arrows add up again, while other funny ways add to a total contribution of zero. One more example: What would happen if light could go several ways, but they would all take the same time? Right, all arrows would point into the same direction and add up to a really big total arrow. Such a setup is what we have in a converging lens! And the large total arrow corresponds to the bright spot of light that we get at the focal point.

In the world of particle physics, the Feynman path calculations also lead to strange effects. As an example, the decay of a Higgs boson (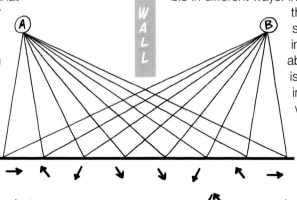[3]) is possible in different ways. Intuitively, you would think that the fact that it has several ways to decay increases the total probability for a decay. But this is wrong. The two most important possibilities will add destructively and cancel to a large extent. So in this case, summing up the two possibilities leads to a total possibility which is lower than each of the two single possibilities. Quantum mechanics is really a strange world, isn't it?

SUM OF ARROWS

↗[3]: *"The Higgs Mechanism" on page 241*

Quantum Tunneling
Where There Is a Wave Function, There's a Way

Quantum particles are really rather strange beasts. One of their unnerving features is that, at any single moment, they are not just at one place. Instead, they are spread out – or rather, what is spread out is the possibility of finding them somewhere when one is looking for them. This spread out possibility is called "wave function", and it's what makes quantum particles weird (↗[1]).

Although one can never restrict it to a single point, the wave function of a particle usually is concentrated at some confined region in space. That is where one is most likely to find the particle, when one is looking for it. But wave functions also have "tails" which spread out far beyond their "core" region. These tails of the wave function tell you that, with a small probability, the particle can be found far away from where one would expect them to be. Sometimes, the particle can even show up at places where it is not supposed to! This is called "tunneling".

Tunneling: Crossing a Barrier That Would Normally Be Forbidden

For instance, a quantum particle can cross a barrier, such as a wall or another kind of obstacle. For this to happen, all one has to do is shoot it directly at the wall. In most cases, the particle will behave in a way one would expect from small, hard spheres: it will simply bounce off the wall. But there is a small chance that it will travel through the barrier as if there was no obstacle at all! How can that be?

Well, as we have said, the quantum particle is actually not a hard sphere – rather, it is described by a wave, which is extended over some region of space. If that wave function approaches an obstacle, like a wall for instance, it will not completely be reflected off it. A small portion of the wave – one of its tails – will be able to reach through the barrier to the other side. Now remember: this wave tells us about the probability where to find the quantum particle. A small part of the wave reaching through the barrier means that there is a small chance that the particle can be found on the other side of the wall!

This chance is actually really small, and it gets exponentially smaller, the thicker the barrier is. So for barriers in our everyday life, such as the door to your flat, the chance of a quantum particle to tunnel through it is really not that big at all. Most of the time, an atom or an electron will just be reflected off it. And the chance that all 10^{27} particles within a macroscopic body (such as yourself) will tunnel through the door is even smaller by a humongous factor.

↗[1]: *"Wave-Particle Duality" on page 143*

In fact, if you have lost the keys to your house, and try to tunnel through the door by running against it – just leave it be. If you tried that once each second, it would take much, much, MUCH longer than the current age of the universe, until there is even a remote chance for you, as a whole, to tunnel though the door. There is a much bigger chance of only half of you making it through the door, the other half staying on the outside. So better just call the locksmith.

So for the regular things of our everyday life, tunneling basically never occurs. A wall is a wall, and things can not magically teleport through it. For single quantum particles, however, tunneling is something that does not only occur reasonably often, but is actually quite important.

Nuclear Fusion: Quantum Tunneling Lets the Sun Shine!

Take the process of nuclear fusion, for example (↗2). In the sun, there are lots and lots of atomic nuclei flying around. Sometimes two of them fuse together into a nucleus of a heavier element. This way a lot of energy is released, which ultimately is the reason for the sun shining. In order for this fusion process to happen, the two atoms need to come together really closely. Unfortunately, because it is really quite hot inside of the sun, they have all been stripped of their electrons. Thus, they are positively charged, and repel each other. Given only the thermal energy, they actually can never come close enough to fuse together. They are simply not hot enough to overcome the electrostatic repulsion.

So one might wonder how fusion can actually happen – the sun is shining, after all! The answer is: tunneling!

The obstacle which the particles tunnel through, in this case, is not a physical wall, but the barrier of electrostatic repulsion. Normally, two nuclei could not come close enough to each other in order to fuse together, because their mutual repulsion is too strong. But, because they are quantum particles, there is a small chance that they tunnel through the electrostatic barrier which keeps them apart. That chance is not very large – but in the sun, there is an abundance of atoms, so that at every moment, there are enough fusion processes happening to

↗2: "Nuclear Fusion" on page 175

keep the sun burning. Lucky for us – the sun is shining because of quantum tunneling.

The STM: Seeing Single Atoms Thanks to Quantum Tunneling

Another example, where the quantum tunneling of particles, specifically electrons, is used for technical purposes, is the so-called "scanning tunneling microscope", or STM. Such a microscope does not have any lenses, and does not use light. Actually, the things one wants to "see" with these are far too tiny for visible light: with an STM one can make individual atoms visible.

The main ingredient of an STM is an extremely fine tip of a needle. This tip is brought very close to the surface which one wants to have a look at. The tip needs to be really, really close – but should not touch the surface. Then, one takes a battery, and connects the surface with one cable, and the needle with another. Since the tip does not touch the surface, the circuit is broken, and there should be no electric current flowing, right?

Actually, wrong: the electrons which make up the electric current have a small chance of tunneling from the tip of the needle to the surface. They can travel normally from the battery to the tip – then there is a small chance for them to tunnel through the barrier, which in this case is just the isolating gap between needle and surface. From there they can travel back to the battery and close the circuit. Although the circuit is broken, there is actually a tiny electric current, because of the quantum nature of electrons.

How can one use this setup as a microscope? Well, remember what we said about the likelihood of a quantum particle actually tunneling through a barrier: That chance depends very sensitively on the thickness of the barrier. So in our case, the further the needle is away from the surface, the smaller the electric current, which one can measure quite easily. So, what one does is to scan the surface with the needle, and closely watch the current. If the current stays the same all the time, then the surface is plain – the distance between it and the needle does not change. If one suddenly sees an increase in current, the distance must have gotten smaller – a sure sign that there is a little bump in the surface.

The most important thing is: this sort of scanning is extremely sensitive, and can register changes in surface height down to the thickness of a single atom! So this is an amazing way to find imperfections in a material, even on the molecular and atomic level. This way, an STM can make very small structures visible, which one could never see with the naked eye.

"Ashes to ashes, dust to dust... unstable particles to radioactive nuclei..."

Radioactive Decay
About the Life of Nuclei and Their End

It gives us fear, when it leads to natural catastrophes. It gives us joy when it produces electricity. No matter what it is doing, it immediately grabs our attention. Its mysterious appearance even names rock songs: radioactivity. There are three things which make radioactivity so interesting. The first is that it leads to a type of radiation which we cannot see (at least not with our eyes) and which harms us. The second is that using it for our benefit (electricity) leads to a type of waste that we have to store for a long time. And finally, the third fact, it tells us that our elements do not last forever, at least not all of them: they decay. This means that a lunch box filled with uranium will no longer be filled with uranium after quite some time.

It might be disturbing that atoms decay. You might ask: "Why is that?" and physicists will answer you: "Why not? Whatever can happen, will happen." Let us take a closer look at an example to see what that means. An atom of a certain element is defined by the number of protons in its nucleus. During the process of a radioactive decay the atomic nucleus changes by emitting radiation, modifying the remaining nucleus. Next to the three characteristic types of radiation, α- (alpha), β- (beta) and γ- (gamma) radiation (↗1) there is also the possibility of a nuclear fission. Let us start with nuclear fission

because you can describe it as "big nucleus breaking apart into two smaller ones", which is quite easy to imagine. Within an atomic nucleus, two different forces are competing: while the strong force (↗2) binds the protons and neutrons together, the electromagnetic force leads to a repulsion of the protons. The strong force dominates the electromagnetic one and keeps the protons and neutrons sticking together. But this amount of domination varies for different nuclei. It might be quite overwhelming, and then a nucleus is very stable and hard to break. But it might also be very slight, and then a nucleus is pretty unstable.

What Causes the (In-)Stabilities

The stability of a nucleus depends on several factors: its volume, surface, number of protons, difference between the numbers of protons and neutrons and the number of proton and neutron pairs. Having all this information, you can calculate the stability of a nucleus using the so-called Bethe–Weizsäcker mass formula (↗3). You might guess it from the name: what it actually spits out is an atomic nucleus' mass, not its stability. But the two are closely related. Let us put a nucleus on a scale and compare its weight to the weight of its constituents, for example a helium nucleus.

↗1: *"Alpha, Beta and Gamma Rays" on page 171*
↗2: *"The Strong Interaction" on page 229*
↗3: https://en.wikipedia.org/wiki/Semi-empirical_mass_formula

Oh wow, the scale is imbalanced! The nucleus is actually lighter. This phenomenon is called "mass defect". It's the $E=mc^2$ game again: part of the protons' and neutrons' masses got converted into energy. This is exactly the energy that is used to bind the protons and neutrons to a nucleus. And also, it is the amount of energy that you need to break the nucleus. So: the larger this mass defect, the more binding energy and the more stable atoms.

These differences in the masses of the nuclei allow for radioactive decays: a nucleus can break into two smaller ones if the sum masses of the "daughter nuclei" is lighter than the mass of the "mother nucleus". The difference in masses is then released as kinetic energy, so the speed at which the daughter nuclei are moving. If you manage to convert this energy into thermal energy you can heat up water and drive a steam turbine with it, which will then run a

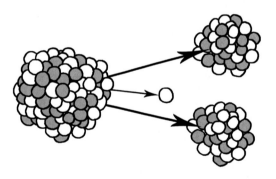

generator and produce electricity. This, simply put, is how a nuclear power plant works.

Nuclear fissions can happen spontaneously. You might well ask: "Okay, if a nucleus can decay into

two lighter ones, why does it not do it immediately?". Once more, the rules of quantum mechanics come into play. Without quantum mechanics, the strong force would keep the nucleus stable. But in quantum mechanics, Heisenberg's Uncertainty Relation (↗4) is valid: The exact position of the nucleus and also of its constituents is not well defined. Within a nucleus, it's two halves (and also all other possible parts of it) – or more correctly, their wave functions (↗5) – are within the nucleus. Mostly. Because a little part of the wave functions is also far outside. This can lead to a point where the two halves are far enough apart that they are beyond the range of the strong force. And here we go: nuclear fission has just occurred. This effect is also known as tunnel effect (↗6).

The Motor of Nuclear Power Plants

Such a fission can happen either spontaneously, or it can be triggered by something else. The most well-known example is uranium-235 (the number tells you the total number of protons and neutrons within the nucleus), hit by a neutron. The neutron is absorbed and the uranium gets pretty unstable and decays into two lighter nuclei, for example barium and krypton (not to be confused with the fictional superman killer kryptonite) and two neutrons. These other neutrons can hit other uranium-235 nuclei and we start from the beginning but two-fold en-

↗4: *"Heisenberg Uncertainty" on page 151*
↗5: *"Wave-Particle Duality" on page 143*
↗6: *"Quantum Tunneling" on page 163*

hanced. We call this a "nuclear chain reaction" and it is the process that heats nuclear power plants.

In this chapter we so far only been talking about nuclear fission. But when people talk about radioactivity, they mostly mean α-, β- and γ-radiation. Still, they base on the same principle for which fission served as a good example: Change into something that is lighter and release the remaining energy into kinetic energy of decay products. We have learned that the time it takes for a nucleus to decay is different. The problem (it's Quantum Mechanics again) is that you don't know when one specific nucleus will decay. You can only give probabilities. What you can do is to define a "half life" as the time it takes until, on average, half of all nuclei of a certain type has decayed.

What to Do With Nuclear Waste?

Half-lives for different radioactive nuclei can vary between less than 10^{-10} seconds and more than 10,000,000 years. Instead of the half-life physicists prefer to define a quantity called "lifetime". It is the time after which only a fraction of 1/2.72 (instead of 1/2 for the half life) still hasn't decayed. The advantage of this time is that you can take the inverse of it to get the activity of a material. If you multiply it with the number of nuclei you have, you know how many of them decay per second. So, very unstable particles lead to a lot of radioactivity. But long lifetimes are also not much better: the activity is lower, but the time that radiation is emitted lasts longer. Unfortunately, nuclear power plants produce radioactive nuclei with a quite a long lifetime. You have to store the nuclear waste at a safe place where its radiation does not harm humanity. One approach is to store the nuclear waste deep underground in former salt domes.

Another approach is the use of so-called nuclear transmutation. With this technique, toxic nuclear waste with a long lifetime can be transformed into less toxic waste or waste with a shorter lifetime, which will then be gone earlier.

Scientists are quite active in the field of transmutation. While it is technically also possible to use transmutation for the artificial production of gold from mercury (unfortunately: economically not efficient), the main focus is on the radioactive waste and the reduction of its storage time by several thousand years. Doesn't that sound encouraging?

Image: German Federal Office for Radiation Protection (Bundesamt für Strahlenschutz)

Alpha, Beta and Gamma Rays
Radioactive Rays

We have learned in the chapter about radioactive decays (↗[1]) that not all atomic nuclei are stable. It is sad that certain nuclei can only be kept for a certain amount of time before they decay. A specialty that comes along with decaying nuclei is that they emit radiation. Two things motivate the emission of radiation. One possibility is that the particle content of the nucleus changes and hence the difference between mother and daughter nuclei has to be emitted via radiation. It is also possible that the particle content does not change, but the nucleus re-arranges it in a way that energy is released.

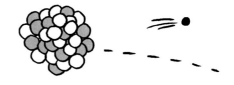

Such radioactive radiation is nothing that people can ignore. You either hate it or love it. And it depends what you want to do with it. Most radiation is ionizing, which means that it has the capability of kicking out electrons from their atoms. This can cause severe damage to the atoms in biological cells, which is why the radiation can be quite dangerous. On the other hand you can also specifically damage "bad cells" from tumors and use the radiation for a medical treatment (↗[2]).

Let us take a look at the three most important types of radioactive radiation. They are named by the first three letters of the Greek alphabet: alpha (α), beta (β) and gamma (γ) radiation.

Alpha radiation consists of little helium nuclei, beta radiation is either a positron or an electron (β[+] and β[-] radiation) and gamma radiation consists of high energetic photons. These types of radiation are normally invisible. But you can use special particle de-

tectors (↗[3]). Let us take a closer look at the different kinds of radiation.

α Radiation

The helium nuclei of the α particles consist of two protons and two neutrons.

Compared to the particles of β and γ radiation, α particles are pretty heavy (about 7300 times heavier than β particles)

↗[1]: *"Particle Decays" on page 221*
↗[2]: *"Radiation Therapy" on page 265*
↗[3]: *"Particle Detectors" on page 253*

and have twice the charge of β particles. Whenever an charged particle – energetic enough to ionize – traverses through matter, it kicks electrons out of the matter's atoms. To do this, the ionizing radiation needs to transfer a certain amount of energy, namely the energy with which the electrons are bound. The more often radiation ionizes, the more energy it loses. The "Bethe–Bloch–Formula", describing a particle's energy loss via ionization, tells us that the loss increases with the particles charge squared. This makes our α particles lose significantly more energy than β particles. γ radiation is not described by the Bethe–Bloch–Formula: even though a photon can also ionize, it will be gone (absorbed

by the particle that it has just ionized) after the first process of ionization.

If you want to shield against α radiation: no problem! As it loses its energy quickly, it can be stopped quickly. A sheet of paper is sufficient. And even the first cells of your skin, which are dead anyhow, stop α radiation. Sounds good? It is. But imagine some radioactive material, emitting α particles, makes it into your body. The α radiation will be stopped pretty fast as well, but this time by your organs. This is very bad: as α particles lose their energy quickly, they deposit a large amount of energy on the short way at which they are stopped and cause cell damage which can lead to cancer. So: better be

aware of α radiation! Surely you would like to know where to find nuclei that do an α decay, not avoid them. One famous candidate is uranium-238. The number after the element's name indicates the total number of nucleons (protons and neutrons). For each element, the number of protons is fixed (92 for uranium) and the number of neutrons can vary. Elements with different numbers of neutrons are called "isotopes". After the emission of an α particle, uranium-238 decays into thorium-234.

You can see: α particles lead to elements with four nucleons less (of which two are protons). If thorium-234 keeps on decaying, it can

end up in radon-222. You can see that only α decays are involved as you can move from 238 to 222 by subtracting multiples of 4. Uranium-238 exists in the crust of our Earth, and so does radon-222. Radon is an invisible gas and it can find its way out of the Earth's crust until it reaches your basement. There it can aggregate and find its way to your lung. This will be the place where radon-222 and its decay products will emit α radiation. It's one of the largest fractions of radioactivity that stresses us humans. So: keep your basements ventilated and don't spend too much time there pursuing your hobby.

β Radiation

During a β decay, an electron or positron is emitted after a neutron in a nucleus changes into a proton or vice versa, respectively. Such a conversion is possible via the weak interaction (↗4) which also leads to an emitted neutrino which we do not see. You see: If one type of nucleon gets converted into another one, the total number of nucleons stays constant during a β decay. A popular example for a nucleus that does a β decay is potassium-40. In most of the cases the potassium will emit an electron (β⁻ radiation), but in 10% it will decay into argon-40 by emitting a positron. You can find lots of potassium in bananas and potatoes, making them a nice source of β⁺ radiation. As these positrons are antimatter (↗5), they will annihilate with their matter counterparts, the electrons, into two photons (↗5,6). This makes radioactive β sources quite popular in medical diagnosis. Depending on which organ you want to screen, you can inject a patient a small amount of a β source. The radioac-

tive nuclei will attach to the organ, produce β radiation and the β radiation will annihilate and produce two photons. If you place a detector around the patient and connect all lines of photon pairs, you can get a nice image of the organ. This method is called PET (positron emission tomography) scan. Even though it is used in medical diagnosis, it is still dangerous. In particular if the energy that the radiation deposits in our bodies gets too high. To shield against β radiation, you need more than a sheet of paper. But a thin plate of aluminum would do it.

γ Radiation

The release of γ radiation is a way for a nucleus to emit energy without changing its particle content. A photon is neither charged nor massive. It is the same particle of which light is made of, but has a lot more energy. You know UV light which has more energy than visible light and can hence burn your skin. While UV light photons have the energy of a few electron volts (eV), X-rays have several thousand eV (kilo eV, keV) and γ photons even millions of eV (mega eV, MeV). γ ray photons do not lose their energy step by step via ionization, as α and β particles do. They rather kick out a single electron off an atom and disappear afterwards. Or they convert into a pair of an electron and a positron (the reverse effect of the electron/positron annihilation). The probability for a γ photon to disappear increases with the thickness of your shielding. This means: the thicker, the better. A γ ray with an energy of 2 MeV will lose 50% of its photons after 1.3 cm of lead, 75% after 2.6 cm and so on. But a few photons will still make it through even a thick lead shield.

↗4: "The Weak Interaction" on page 233
↗5: "Antimatter" on page 217
↗6: "Feynman Diagrams" on page 225

"How much energy am I getting out of it?
Well... not exactly 'stellar'..."

Nuclear Fusion
Energy Source for Dinosaurs and Future Humans

The typical futuristic power plant of science fiction movies, books and videos games is a nuclear fusion reactor. It sounds quite impressive, but what is actually the story behind this nuclear fusion? Does it really exist? It does!

Good Combinations That Bring You Energy

Nuclear fusion might indeed be the optimal source of energy for the future of mankind. But it also used to be a source of energy for the dinosaurs. And it still is for us. But the problem is that we cannot control, or use it in a power plant. We have to use the nuclear fusion that nature provides. You might wonder what that is. It is something we can see very day, but not every night. Right, the sun! Within the sun, an incredibly large amount of atomic nuclei fuse to bigger ones. And by that, they release energy. How can that be? Let us start with the smallest atomic nuclei we know: hydrogen nuclei. These consist of a single proton. If you bring two protons very close together, they can combine to a bigger nucleus: a deuteron, the nucleus of deuterium atoms. Deuterium consists of an electron and a proton (just like hydrogen) plus an extra neutron in the nucleus. But the electrically neutral neutron does not change the atom's chemical properties. It's like hydrogen, but twice as heavy. But wait: We started with two protons and now we have a proton and a neutron within the deuteron. How can that be? One

of the protons must have converted to a neutron! This weak interaction (↗[1]) is responsible for that. Next to the weak interaction you need the strong interaction (↗[2]) to do its work, namely to make the proton and the neutron stick together.

When we said that the mass of the deuteron is twice the mass of a proton, we were not 100% correct. It is 1.999 times the mass of a proton. But where did the rest go? Even if the rest is a small amount, we cannot neglect it. Part of this rest was converted via the weak interaction into a positron and a neutrino. This is needed to keep the charges conserved (↗[3]). But there is still something missing, and the amount that was leftover was converted into energy. In the same way as for nuclear fission reactions (↗[4]) (parts of) masses are converted into energy. We know this as "mass defect" (↗[5]). So we see that you can create energy out of mass by either nuclear fission reactions or via nuclear fusion reactions. If you think you can get an infinite amount of energy by letting a nucleus decay via fission, gain energy, let the fission products fuse again, gain energy and so on … that is not possible. Only light nuclei release energy in a fusion process and only heavy ones release energy via fission.

Cross the Barrier – Cheating Allowed

While some heavy nuclei automatically decay via fission, others need an external trigger like a neu-

↗[1]: "The Weak Interaction" on page 233
↗[2]: "The Strong Interaction" on page 229
↗[3]: "Conservation Laws" on page 51
↗[4]: "Radioactive Decay" on page 167
↗[5]: "E=mc²" on page 237

tron that hits the nucleus and splits it. For fusion, the situation is slightly more complicated. Think of the two protons.

To make them fuse we need two inter-actions: the strong and the weak interaction. But both have a range of only about a femtometer, so 10^{-15} m. So this is how close you have to bring two protons together. The problem: before the strong and the weak interaction can start to act, the elec-tromagnetic interaction will do its job, as its range is infinite. As both protons are positive, the elec-tromagnetic interaction will cause a repulsive force. Even worse: the correspond-ing force caused by the Coulomb interaction increases with the in-verse of the squared distance. So: super short distances, super strong repulsive forces! We see this illustrated: The pro-tons will need more and more energy to come closer and closer. It is like running up a hill which gets steeper and steeper.

You know it from cycling with your bike up the hill: it can be tough. And the higher you want to go, the more energy you need. So somebody has to give the protons a lot of energy to let them fuse. If we give the hint that temperature is just another mea-sure of kinetic energy of the particles in a gas, can you think of a place with a lot of high energetic par-ticles? Right again – the sun!

And here we go: the biggest nuclear fusion reactor for both the dinosaurs and us is the sun. Within the sun, an incredibly large amount of fusion processes take place. The sun is nothing else than a big gas ball of mostly hydrogen atoms. And as it is so big, gravitation can hold it all together to a big ball. The protons in the center of the sun can really feel the pressure of the sun's own gravita-tion. It compresses the hydrogen gas and heats it. This heat allows the protons in the sun be very fast and hence to get very close. Very, very close. But not close enough to let the strong and weak inter-action act. The protons need to do a trick: they use the tunnel effect (↗6). Quantum mechanics tells us that even if a particle does not have sufficient energy to cross a barrier, its wave-like properties allow it to simply tunnel through it. The probability is not too high, but still larger than zero. And only due to this tunneling the protons in the sun fuse. And the fusion does not end at the deuterons. Adding an-other proton to a deuteron leads to helium-3 (two pro-tons and a neutron) and two of such helium-3 nuclei fuse to he-lium-4 (with two neutrons) plus two released protons. And these two protons can then again … let the story of fusion be-gin. This process is called the proton-proton chain reaction. In heavier stars and during supernova ex-plosions, helium and even heavier nuclei can fuse and produce elements up to iron.

A Sun on Earth?

Doesn't it sound nice: While nuclear fission reactors need the expensive, rare and radiating uranium,

↗6: *"Quantum Tunneling" on page 163*

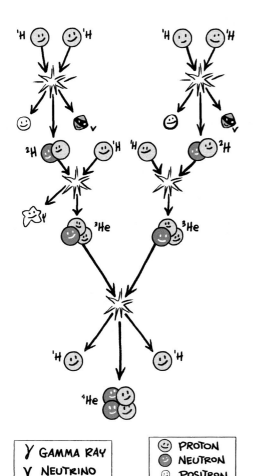

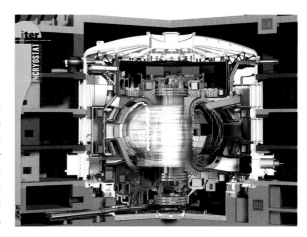

nuclear fusion reactors would only need hydrogen (cheap, plenty on Earth). So it seems tempting to build something like a mini-sun down on Earth and use it to produce energy. There are already many attempts in this direction. What needs to be done in order to run a fusion reactor is to heat up hydrogen (there are also attempts to start with deuterium and tritium) up to several million degrees (here, it does not matter if we use Celsius or Fahrenheit). One also needs to keep this heated hydrogen, which is called plasma because electrons are separated from the protons, in a stable and dense environment. There are several options to heat the plasma: microwaves or fast neutrons can be shot on it or an electric current can be induced within the plasma. The last option you might have seen in a much milder way when you observe electric cables getting warm when they transport a high current. Keeping the plasma together is reached via magnetic fields. Very strong and precise magnetic fields are needed to keep the plasma dense and stable. Several of such fusion reactions are currently tested (↗7), but so far all of them consume more energy for heating the plasma than they produce via fusion. The plan for the future is that the situation is reversed. Let us end the chapter with an image of the ITER reactor, an international project which is currently being built in Cadarache, France. It is of a donut shaped Tokamak type. Other concepts are also tested, such as a Stellarator type or laser-based inertial confinement fusion. Let us see what the future brings!

↗7: https://www.ipp.mpg.de/16900/w7x *and* http://www.iter.org
Image: © ITER Organization

"They give me height, but they make my feet cold."

Superconductors
Super Highways for Electrons

Did you ever ask yourself why electric power lines have such enormously large voltages (several thousands of volts) compared to the voltages of 120 V (US) / 230 V (Europe)? To understand why this is so we have to think about the way that electricity is transported in a cable. Electric charges are moving, accelerated by the electric field inside the cable that appears once you apply a voltage to it. The movement of these charges is the electric current. You can think of such a current as cars on a highway. A high current corresponds to either a lot of cars or cars with a high speed. But there is – fortunately – one difference between the movement of cars on a highway and electrons in a cable. While the cars try not to interact with either each other or the traffic barriers, the electrons do. The amount of that interaction limits the current and is quantified as electrical or Ohmic resistance. Ohm's law states exactly that: For a certain voltage U the current I is limited by the resistance R, by the formula U = R · I. During the interaction between the electrons from the current and the electrons from the conductor in which they are moving, they transfer energy. We get a conversion from electric energy to thermal energy ($\nearrow$[1]) and the conductor warms up. You can verify that, if you use a device with a large energy consumption that needs a large current, let's say a water boiler. Touch the cable! It's warm, isn't it?

Ohmic resistance leads to unwanted effects. Power plants produce electricity, transfer it to you and on the way, part of the energy gets lost as the transfer cables heat up. A few percent of the transferred

power gets lost on the way. This loss is on the one hand larger for higher resistances of the cable. On the other hand it decreases for higher voltages.

This is the reason for the high voltages of electric power lines. It is clear that running with higher voltages is not always an option for decreasing power losses. Ohmic resistances limit, for example, the current we can use in electromagnets. And as we get stronger magnetic fields for higher currents, this also limits the magnetic field strengths that we can get.

A good question to ask now would be: Is there a way to reduce this Ohmic resistance? There is. It depends on the thickness and the length of a cable. The larger the cross section and the shorter the cable, the lower the resistance. But short and thick cables are often not very practical. There is another dependence of the resistivity: the temperature of the conductor. While for a few conductors the resistance decreases for higher temperatures (mostly semiconductors), for most of them it increases.

The Unexpected Jump
in Supercold Regions

In 1911, the Dutch physicist Heike Kamerlingh Onnes measured the resistance of mercury. He varied the temperature and checked the principle: the colder the temperature, the lower the resistance. Liquid helium can be used to cool things down

$\nearrow$[1]: *"Conservation Laws" on page 51*

to very, very low temperatures. So he measured the resistivity at 4.4 K (Kelvin), which corresponds to -268.75 °C (Celsius) or -451.75 °F (Fahrenheit). The Kelvin scale is used as it has a well-defined value of 0: there is no way to get any colder. Temperature corresponds to the movement of atoms, and at 0 K they all stand still. So we have very, very cold mercury. And a quite low resistance. Kamerlingh Onnes went down to 4.3 K: even less resistance. But then, suddenly, at 4.2 K, there was a big jump and the resistance went down to 0. No Ohmic resistance. This was totally surprising.

What he had just discovered, and for which he was awarded a Nobel prize in 1913 (he got it not only for this observation, but for his work with extracold temperatures in general, see also ↗[2]), was the phenomenon of superconductivity. It is defined as electric conductivity without resistance. The phenomenon was a total surprise for the whole physics community. And it took about 50 years to find an explanation for this effect. In 1957, the physicists Bardeen, Cooper and Schrieffer (BCS) formulated a theory which tells us what is actually happening. Simply put, an electron from the current attracts positive charges from the conductor. A second electron from the conductor joins it and the two electrons form a so-called "Cooper pair". While a single electron is a fermion (with a spin of 1/2), of which at most one with the same properties can be at the same place, a Cooper pair is a boson (with integer spin): the two electrons' spins of 1/2 point into either same or opposite directions and lead to a total spin of

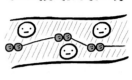
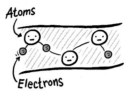

NORMAL CONDUCTOR
Atoms
Electrons

SUPERCONDUCTORS

MAGNETIC FIELD LINES TRAVELING THROUGH A CONDUCTOR

1 or 0. The good thing about bosons: You can put as many bosons with the same state at the same place as you like (↗[3]). So all of these Cooper pairs will take the same state and act like a whole. In case one of the conductor's atoms wants to interact with an electron with the current (as it happens in the case of Ohmic resistance), it has to interact with all of them at once. And at such low temperatures, the energy of the conductor atoms is not sufficient. This is the reason why the electrons can pass without any resistance.

Shoo, Magnetic Field!

A nice effect comes along with superconductivity. Usually, if you put a conductor into a magnetic field, the magnetic field lines go their way through the conductor. The physicists Walther Meissner and Robert Ochsenfeld observed that in contrast to normal conductors, superconductors expel the whole magnetic field. This is caused by currents induced by the magnetic field at the surface of the superconductor. These currents induce a counterfield which cancels the external magnetic field in the

MAGNETIC FIELD LINES PUSHED AWAY FROM A SUPERCONDUCTOR

SUPER

↗[2]: *"Superfluidity" on page 183*
↗[3]: *"Spin" on page 187*

conductor. And here we go: a field-free conductor. This effect is not only interesting for a physicist, but also nice to watch. A superconductor, expelling all magnetic field lines, will act as a perfect diamagnet (a magnet that, once put into an external magnetic field, will induce its own field in the opposite direction and will be repelled by the external field) and will levitate if placed in an external magnetic field as you see on the photo.

New Possibilities

Imagine the possibilities of superconductors! No more electric resistance means that you can transport energy without losses. While the cooling of the superconductors is still quite complex and expensive, some electricity providers have already started using superconducting cables. If the cables no longer heat up, you can also transport a lot more power through them. Depending on the model, household fuses tend to blow out when a current of 20 amps flows through (to prevent the cables from melting) a superconducting cable has no problem with even much larger currents. In the second photo on this page you can see two cables, both ca-

pable of transporting 12,000 amps. The big one is a classical conductor that needs to be cooled with water in the inside. The small one is a superconductor. High currents in thin cables are needed to produce very high magnetic field strengths in electromagnets. So superconductors got very popular whenever strong magnetic fields were needed, for example in particle accelerators (↗4) or MRI (magnetic resonance imaging).

Physicists investigate different materials that can be used as superconductors. In the magnets of the particle accelerator LHC (the thin cable in the photo) a material consisting of niobium and titanium was used. The temperature at which materials become superconducting varies. Of particular interest are those that do not have to be cooled to 4 K, as for the LHC, but only to about 100 K (which is still -280 °F, but hey…). This allows using liquid nitrogen instead of liquid helium (which is rare and really expensive). The hope is that in the future, new materials will be developed, which need even less cooling. But it is clear by now that such materials will be very complex and hard to produce.

↗4: *"Particle Accelerators"* on page 249
Image (upper left): © CERN
Image (lower right): Boris Lemmer

Superfluidity
The Creepy Kind of Fluid

Let us talk about fluids. The term "fluid" describes both liquids and gases. We learned about the way that fluids move (↗[1]). We also got to know a type of fluid with very special properties, the non-Newtonian fluid (↗[2]). While the non-Newtonian fluid was something in between a fluid and a solid state, this time we will talk about something that is fluid for sure.

What is it that actually defines a fluid? Matter, such as water, exists in different phases: solid, liquid, gaseous or as a plasma. The actual phase depends on the pressure and the temperature. These phases are caused by forces that act between the different molecules of a type of matter. Water molecules can attract each other and form a solid crystal. Solid matter can neither easily change its form nor its volume. But as soon as the water molecule's average energy (and this is what we define as temperature) is too large, it breaks the bindings and the molecules can move. That's why you can deform liquid water. But the water molecules are still packed closely, that's why you cannot compress it. If you increase the temperature further, the molecules will no longer stay together closely, they will be free. And as this distance can be decreased by applying an external pressure, you can compress a gas. If

you further increase the temperature, you will even rip off the electrons from the rest of the atoms. This state is called a plasma. You can read more about such phase transitions in the chapter about the childhood of the universe (↗[3]).

Properties of Fluids

Getting back to fluids: You could classify them, next to their elementary properties, by their behavior. Fluids conduct heat, for instance. Some do more, some do less. This effect can be quantified via the "thermal conductivity" λ. Sometimes you want to minimize it, for example to keep your house warm during winter. Then you pick materials with a small λ, which do not transport the heat outside. The lowest thermal conductivity you can think of has a vacuum. There, λ is 0. But there are also cases where you want materials with a large λ, for example to transport heat away quickly to prevent damages to machines which get heated up. Copper and silver for example have values of λ about 400 Watts per meter and per second.

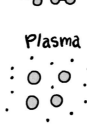

Next to the thermal conductivity, you might be interested in a fluid's viscosity. The viscosity tells you "how thick your fluid is". The more viscous a fluid is, the slower it moves. A physically more precise defi-

---

↗[1]: *"Fluid Flow and Turbulences" on page 27*
↗[2]: *"Non-Newtonian Fluid" on page 39*
↗[3]: *"Timeline of Our Universe" on page 100*

nition than "thick" is that viscosity defines the resistance against shear stress. The illustration tries to explain it. If you keep the lower plate fixed and move the upper one, the uppermost fluid layer will move as well, while the lowest will not. The viscosity tells you how much shear force per area you have to apply to reach a certain velocity for a given distance between the plates. For a great example of this, think about how water and honey behave. A funny side remark: If you are looking for something with a really large viscosity, take pitch! An experiment, started in 1927, is measuring the rate at which pitch drops out of a funnel. Since the beginning of the experiment, only nine drops made it. And only the last falling drop was recorded via web-cam. This experiment, which is still running, made it into the Guiness book of world records as the longest lasting lab experiment on Earth.

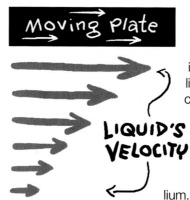

Helium – Cold, Colder, Strange

But back to fluids! Let us talk about a special one: helium. Helium atoms consist of two protons and two neutrons in the nucleus as well as two electrons. We find a lot of helium inside the sun, as it is the product of nuclear fusion (↗4), the process that heats it.

The helium that we can find on Earth does not originate from the sun. It comes from decaying radioactive elements inside the Earth's crust, such as uranium. During such radioactive decays, a parti-

cles are emitted (↗5). These are already the nuclei of helium atoms and just need to capture two more electrons from the surrounding Earth crust. You can extract helium from natural gas. It is used for a variety of things: If you inhale it, you will sound like Mickey Mouse as the speed of sound in helium is about three times larger than in air. As its density is also lower than air, it is a popular balloon gas. Helium balloons can fly, but they cannot burn. That is actually its big advantage compared to hydrogen, which could also make balloons fly but is much more reactive (and hence flammable) than helium. The most popular usage of helium, however, is cooling. If you've ever seen an MRI scanner in a hospital: the coils of its electromagnets are cooled with helium. Also, the world's biggest particle accelerator (↗6) is cooled with helium.

Speaking of cooling: what does actually happen if you cool helium? At the beginning of the chapter we talked about phase transitions. So sooner or later during cooling, gaseous helium should become liquid. This transition is called condensation. This does indeed happen, but while our good old water condensates below 212 °F, or 373 degrees Kelvin, helium will remain gaseous down to 4 degrees Kelvin! You see: there is not much room down to absolute zero for helium to become solid. Indeed, it has to be colder than 1 degree Kelvin. But at normal conditions it will not turn solid, only if you increase the pressure to 25 times

---

↗4: *"Nuclear Fusion" on page 175*
↗5: *"Alpha, Beta and Gamma Rays" on page 171*
↗6: *"Particle Accelerators" on page 249*

the atmospheric pressure. Surely, this is quite impressive. But it isn't a good enough reason to devote a whole chapter to cold helium.

Supercold, Superfluid

In the chapter about superconductivity (↗7) we learned that certain materials will turn into a special quantum state when cooled below a certain temperature: they will become superconducting. This means that they lose all their electric resistance. If helium gets colder than 2 degrees Kelvin, it turns into a so-called "superfluid". Similar as in superconductors all electron pairs behave as a single particle, all helium atoms of a superfluid can be described by a single wave function (↗8). "So what?", you might think. But this leads to very interesting changes in helium's properties.

The thermal conductivity describes how fast temperature can be transported. The hot part of a material has atoms which move fast. By colliding with their neighboring atoms, they transfer their velocity to them. Now imagine shaking a single wave (function) at one end. The reaction on the other side will follow immediately. Think of a stiff wave that you shake on one end. The reaction on the other end will follow immediately. So heat, represented by movements of the matter's atoms, will immediately be transported from one end to another. That's why the thermal conductivity of a superfluid is infinite!

What about a superfluid's viscosity? While the friction of electrons in a conductor leads to electric resistance, the friction of layers in a fluid leads to a viscosity. And the same way as the electric resistance disappears completely, superfluids suddenly get rid of their viscosity. Imagine viscosity-less honey on your spoon. How long could you balance it?

But it's not only the balancing on a spoon that gets pretty tough for a superfluid. Even if you put it into a cup and place it on a table, it will not stay there. The "Onnes effect", named after the physicist Heike Kamerlingh Onnes, describes what will happen: The superfluid will crawl up the cup and crawl down to the ground, as we have illustrated it. While this is impossible for normal fluids, the superfluid can crawl upwards as the forces with which the fluid sticks to the wall are larger than the gravitational force. And as there is no more friction within the fluid (no viscosity), a thin film of the superfluid will cover the cup and make its way to the place of lowest potential energy (as a ball rolls down a hill), which is down the table.

The Onnes effect also makes it very difficult to store ultracold liquid helium. Wherever there is just a very small hole in the helium vessel: it will always find its way out!

SUPERFLUID

CUP

↗7: "Superconductors" on page 179
↗8: "Wave-Particle Duality" on page 143

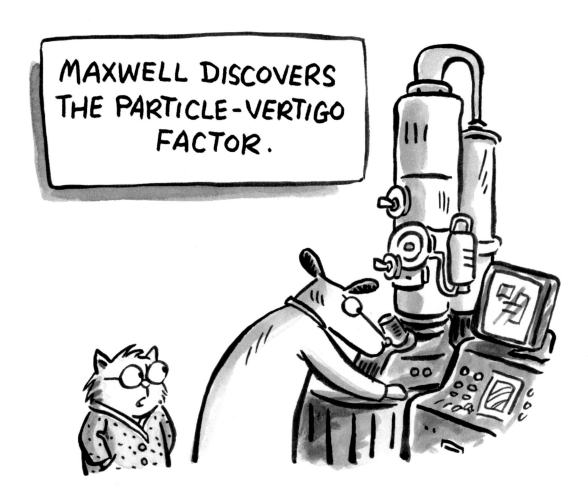

"What do you mean an electron threw up on the condenser lens?"

Spin
Particle Dances in Discrete Steps

In everyday life, magnets seem very mysterious. They exert a force (onto other magnetic materials), which seems to come from nowhere. This allows you to make amazing fridge poetry sticking to your favorite piece of kitchen furniture, seemingly forever. Where does this force come from?

Magnets – This Is How They Work

Well, it comes from the magnetic field that magnets produce (↗[1]). Just as electric fields, magnetic fields have a direction and a strength. They are what's called a vector field. The direction is what is telling a magnet into which direction its North Pole should face. As soon as it faces the right direction, it begins to wander towards the direction in which the magnetic field is strongest.

Okay, but how does one actually produce a magnetic field? What is it, what makes a magnet into a magnet? Well, this has been found out by James C. Maxwell, about 150 years ago: magnetic fields are produced by accelerated charges. So whenever a charged particle changes its direction, anywhere, it produces a magnetic field which exists as long as the change takes place. To produce a constant magnetic field, the best way is therefore to make a charged particle to fly in circles. A great example, which you

might have seen in real life, are copper solenoids: electric coils which are used to generate magnetic fields. But electrons flying in circles is also happening in almost all atoms: an electron flying around an atomic nucleus is perfect for this! Even though quantum mechanics is a bit weird, and one cannot conceivably say that an electron "flies in circles around a nucleus", it is doing something quantum mechanically which comes pretty close to this. And indeed, many single atoms behave like little magnets because of this.

So is that what is making magnets into magnets? All the electrons in the orbital shells of their atoms, dancing in unison to make one giant, big magnetic field? As it turns out, it is not that simple, because in most materials the electrons cannot just circle the atoms in the direction they want – rather they are used to bind several atoms together, and this job requires very specific movement from the electrons. One that does not allow for all of them to fly in circles the same way.

But that's no problem: many materials still behave like magnets, because of the magnetic fields generated by their electrons. How does that happen? Well, as it turns out, a single electron still produces a magnetic field of a certain strength, even

↗[1]: "Light" on page 7

when it does not change its direction. It does that even if the electron does not move at all!

Spinning Electrons Are Magnets as Well

When physicists realized this, over a hundred years ago, they were quite puzzled, because that seemed to contradict Maxwell's theory. But they came up with a solution: The electron's intrinsic magnetic field generated by the electron behaved as if it was generated by the electron spinning around itself, with a very specific rotational velocity. If the electron were a tiny charged ball, that would in fact, by Maxwell's theory, generate a magnetic field. That is why this physical quantity is, up to this day, called spin. Nowadays we know that electrons are not little charged balls, but are more complicated objects, with quantum weirdness all over them. But the term was so catchy, and the idea of the electrons as little spinning balls was so appealing, that it stuck in the minds of generations of physicists. Even today, one can find this image in many physics textbooks, although it is an oversimplification of what is actually going on with the electron.

The Spin – More than Just a Number

The strength of the generated magnetic field (related to the bona fide "spinning velocity"), has the same value for all electrons. In Planck units, it has the value of one half. Later physicists have found that all elementary particles have a spin, and these come only in steps of one half. So a particle can either have a spin of zero (such as the Higgs boson, ↗²), one half (such as all the quarks, the electrons and neutrinos), one (the interaction bosons like gluons, W- and Z-bosons and the photon – which does not generate a magnetic field itself because it is not charged). Currently, nobody has found an elementary particle with spin three halves (but there are some composite particles, consisting of several quarks, which have that spin). The same is true for spin two, but if there is something like a graviton, the alleged boson for the gravitational interaction, it should have spin two.

I A·A·M A M·MA·AGNET!!

A SPINNING ELECTRON

Bosons and Fermions

As you might have noted, particles with integer number spin are called bosons. The one with half-integer spin also have a dedicated name: they are called fermions. This distinction is quite important: bosons and fermions behave fundamentally very different. It all boils down to the question of "how many particles can I put into the same state?". In other words: how many particles of one type can be made identical, so that all their physical properties are the same?

↗²: "The Higgs Mechanism" on page 241

The answer for bosons is: as many as you like! You can pile as many bosons on top of each other, then they have the same position, velocity, and other properties as well. That is what is commonly referred to as Bose-Einstein condensate (BEC). In a BEC, all bosons are exactly the same – which is why it behaves like one, giant particle. BECs have been quite fashionable in research in recent years, by the way! BECs can result in phenomena like superconductivity ($\nearrow$3) and superfluidity ($\nearrow$4).

The answer for fermions is: at most one! That means that fermions behave much more like "solid balls" than bosons. If one fermion already has a certain position, velocity, spin, etc., then every other fermion of the same type in the universe must differ from the first one in at least one quantity. It must be somewhere else, or if it is at the same place, it at least must have a different velocity. If both are at rest and at the same place, their spin must point in different directions.

Quantized Spin: Either with Us or against Us

Speaking of the spin – there is another peculiar property, which demonstrates the weirdness of quantum physics. The spin of an electron is like a little arrow: the length of this arrow tells you how strong the magnetic field is that the electron generates. The direction of this arrow is where its North Pole points. You can think of this as the "spinning axis" of the electron.

The arrow length is the same for all electrons, as we have already said: it is one half (in Planck units). The direction, however, is not arbitrary at all. If you place an electron in an external magnetic field, it behaves like all magnets – it wants to point into the direction of the magnetic field lines. But an electron does not turn around gradually, like normal magnetic needles in a compass would. Rather, it can either point in the direction of the magnetic field, or it can point in the opposite direction. But any other direction in-between is not allowed.

This is very similar to the way energy can be exchanged by the electromagnetic field: energy change can only happen in certain steps, not gradually ($\nearrow$1). Spin is also something that can only change in steps of one Planck unit. Either the spin points along the magnetic field lines (one then says the spin is $+1/2$), or in the opposite direction (then the spin is $-1/2$). And since the difference between $+1/2$ and $-1/2$ is precisely one (Planck unit), the electron can only switch between the two possibilities, but cannot point into any other direction, say, sideways.

This is not the whole story, however. An electron's spin can not point into any other direction than along or opposite the magnetic field. But, being a quantum particle, it can also be in a superposition – pointing into both directions at the same time ($\nearrow$5), with various different possibilities. This makes it possible to build quantum computers ($\nearrow$6), at least in principle.

$\nearrow$3: *"Superconductors" on page 179*
$\nearrow$4: *"Superfluidity" on page 183*
$\nearrow$5: *"Schrödinger's Cat" on page 155*

$\nearrow$6: *"Qubits" on page 199*

Entanglement
A Spooky Action at a Distance?

As we have stated in many articles, the quantum world is weird. For very small things, the logic of everyday life just does not apply any more. Where a classical thing has to be somewhere, but nowhere else, a quantum thing can be in several places at once, with different probabilities.

Electron Spins: Up, down, or Both!

One great example for this is the spin of a particle (↗[1]), most notably that of the electron. The spin is what makes the electron into a little magnet. In an external magnetic field there are only two possibilities for it to align itself: either with the magnetic field lines, or against them. But, an electron being a quantum particle, it can also be in a superposition of these two possibilities (↗[2]). Just as Schrödinger's cat can be both alive and dead at the same time, so can the spin of the electron point both towards, and opposite to the magnetic field. In terms of the "spinning charged ball" picture of the electron, this means that the electron is spinning both clockwise and anti-clockwise at the same time!

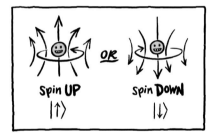

Now not only is each electron a quantum system, and can therefore be in superposition of two states, two (or more) electrons can be as well. This means that the physical system "two electrons" can be in superposition of two possibilities. This is in fact much more interesting than just one electron, because this means that the two electrons can be entangled.

Entangled Electrons: Spin, Baby, Spin!

To describe the prime example of two entangled electrons, consider them both in a magnetic field, which has field lines pointing upwards. So the spins of the electrons are pointing either up or down. Even better, the system of the two particles can be in the superposition of the following two states: the first electron has "spin up" and the second one has "spin down" – that's the first state – and the first electron has "spin down" and the second particle has "spin up" – that's the second state.

Just as with Schrödinger's cat

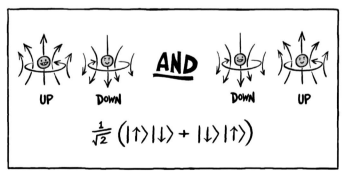

↗[1]: *"Spin" on page 187*
↗[2]: *"Schrödinger's Cat" on page 155*

(↗²), the two electrons stay in this superposition, until it is disturbed – and measuring the spin of one particle counts as such a disturbance. Imagine now someone could ask the question "Is the spin of the first electron pointing up or down?", and perform a measurement to find out. Then, by the rules of quantum mechanics, there is a 50-50 chance of finding that the spin is pointing up, and a 50-50 chance to find that it points down. The spin of the second electron is then completely determined, without looking at it: it is down or up, depending on the result of the first measurement. The spin of the first electron is random up or down, but after measuring it, one knows the second spin is immediately fixed to be the opposite.

The Spooky Action at a Distance

Now that does not seem to be too surprising at first glance. If one plays the Shell game, with two cups and one marble, then the marble is under one cup, but not the other. If one lifts one cup and finds the marble, one knows immediately that there is no marble under the second cup, without lifting it up and looking. But that is not the situation with quantum mechanics, because in the Shell game, the marble was already under one cup, but not the other, even before we lifted one up and looked. We just didn't know which situation we were in. In the case of the two entangled electrons however, it is decided at the moment of the mea-

surement, whether the spin of the first electron is up or down. There was no missing information.

The Einstein–Podolsky– Rosen Paradox

Okay, you might say, then the measurement of the first electron changes its state, and somehow the first electron tells the second one that it was just measured, and the second electron flips its spin accordingly. But that is precisely the point which made Einstein, Podolsky and Rosen very uneasy, around a hundred years ago. You see, the two spins of the electrons can be entangled, and then the two electrons can be moved away from each other, arbitrarily far (if you're careful enough)! Einstein, Podolsky and Rosen argued: so what if the distance of the two electrons is, say, one light year when the measurement of the first electron is performed. The flip of the second spin to the opposite direction of the first spin happens in that moment in which the first electron's spin is measured. So was the information which spin had been measured transmitted faster than light between the two electrons? Understandably, in particular Einstein was not happy about this, because he had just founded his career on a theory which predicted that nothing could travel faster than light (↗³)!

This so-called EPR (after the name of its inventors) paradox has puzzled many physicists, and made

---

↗³: *"Relative Space and Time" on page 117*

Einstein talk about a "spooky action at a distance" in quantum mechanics. He absolutely didn't like it. But people have come to realize that it poses no actual problem. So although, for all intents and purposes, the second electron changes its spin in the same instant the measurement is performed on the first electron a light year away, it is not possible to transmit information with this process.

No Communication Faster than Light: Star Trek Thwarted again!

Here is what would happen: Assume you create lots and lots of pairs of electrons with entangled spins, and give the first bunch to a physicist here on Earth, and the second one to a physicist somewhere far away, say the Andromeda galaxy. These two physicists then agree to start measuring the spins of these electrons, in a pre-determined order, at the same day, at the same time on the clock. What they will both find is, about half the electrons will have

sa. But in order to find out that piece of information, they first need to meet and compare their results. And that can happen only slower than light. So in other words, whenever one of the two physicists measures a spin, there is no way of determining whether that electron had still been entangled at that point, and whether she was just fixing the spin for the other physicist far away, or whether the other physicist had already measured the entangled pair, and she was just measuring the opposite result of the other physicist's measurement. And it is good that you cannot determine the difference between the two, because Einstein's relativity theory tells you that these two situations are actually indistinguishable: a third observer looking at the two physicists (and wondering whether that's what they always do for fun on a Saturday afternoon) will see either one or the other situation taking place, depending on how fast and which direction he flies. So relativity is protected from the spooky action at a distance of quantum mechanics.

spin up, and the other half will have spin down. Both will conclude that the result of the measurement will be completely random. It is only after one of them has hopped into a spaceship and traveled to the other to compare results, that they will find that, whenever one has measured "up", the other one has, infallibly, measured "down", and vice ver-

By the way: although its sounds very mysterious, quantum entanglement is something that is produced and measured in laboratories around the world every day! And physicists, in fact, don't do this just for fun, but rather because entanglement is the core mechanism behind quantum computing (↗4) and teleportation (↗5).

↗4: *"Qubits" on page 199*
↗5: *"Quantum Teleportation" on page 195*

"Sigh. Maybe someday."

Quantum Teleportation
There and Back Again

When one hears about the weirdness of the quantum world, one gets the impression that it is mostly about fuzziness and uncertainty. It seems that, as soon as we want to know anything specific about quantum particles, we hit a brick wall. Want to know which of the two slits the electron traveled through? No can do (↗[1]). Want to know position and momentum of an electron at the same time? Bad luck (↗[2]). Is the cat in that box dead or alive? Probably both (↗[3]). We don't seem to have a good grasp of the quantum world.

The Quantum World: Mysterious …

But that impression is wrong! We know precisely how the quantum world works – it's just that it works differently from the world we experience every day. There are certain questions we cannot expect to get answers to, because they are nonsensical questions. A quantum particle does not have a well-defined position and momentum at the same time. Asking where and how fast a particle is precisely is like asking how the color green tastes.

We do not only have a good idea of how the strange world of quantum mechanics works, we can also use its properties to our advantage, to accomplish amazing things! One of the most interesting examples is quantum teleportation. One can use the fact that particles can be entan-gled (↗[4]), together with the fact that every measurement of a system changes the system, to teleport a qubit (↗[5]) from one place to the other.

… But We Can Bend It to Our Will!

Now, before you get your hopes up for teleporters like in Star Trek – that is not quite what we are talking about. What we mean by quantum teleportation, is that a spin state of an electron – a qubit – can be transferred from one electron to another, without moving the electrons themselves.

Emmy's Got an Electron

Imagine that, in her lab, Emmy has an electron with a certain qubit. That means, its spin could be up, down, or any quantum superposition in-between. Actually, Emmy will most likely not know which state the electron is in precisely. She doesn't need to, for the teleportation to work. Actually, if the qubit is unknown, there is no way of finding out what it is precisely, either. One could try to measure it, but the only results one can get are "up" and "down". That in itself would not tell us much about which qubit the electron was in, precisely – and after that measurement we would have changed the spin anyway, so there would be no way of finding out!

But this is the marvel of quantum teleportation: the qubit can be transported

---

↗[1]: *"The Double Slit Experiment" on page 147*
↗[2]: *"Heisenberg Uncertainty" on page 151*
↗[3]: *"Schrödinger's Cat" on page 155*
↗[4]: *"Entanglement" on page 191*
↗[5]: *"Qubits" on page 199*

without knowing it! You never measure the qubit – you actually measure something different.

Maxwell Awaits a Message from Emmy …

Now, there are going to be more electrons involved in the whole teleportation process, so we'll start giving them names.

Maxwell sits in another lab. Before the whole experiment has been set up, Emmy and Maxwell have prepared something: they have created a pair of entangled electrons (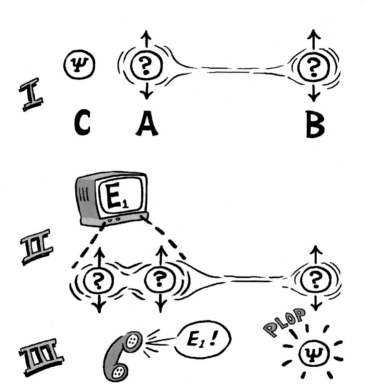⁴). Let's call them A and B. Emmy has taken electron A with her, and Maxwell has taken electron B with him. They have to handle them carefully, so as to not destroy the entanglement between them.

Emmy is in her lab, and has her original electron – let's call that one C – which has some unknown qubit, which she wants to teleport to Maxwell. She also has electron A, which is entangled with the one Maxwell has in his lab. Now the trick comes: Emmy performs a measurement on A and C at the same time. She measures how much the two are entangled.

Wait a minute – weren't A and B entangled? C wasn't entangled with either of them! Well, that is right, but remember that every measurement process changes the state of the system. After Emmy has measured the entanglement of A and C, they will be entangled in some way afterwards! And her measurement will tell her precisely in what way.

This measurement process by Emmy has some influence on B as well, though. As it turns out, because A and B were entangled, the measurement of A and C changes the state of B, too. In fact, after Emmy determined the entanglement between A and C, Maxwell's electron B will not be entangled anymore

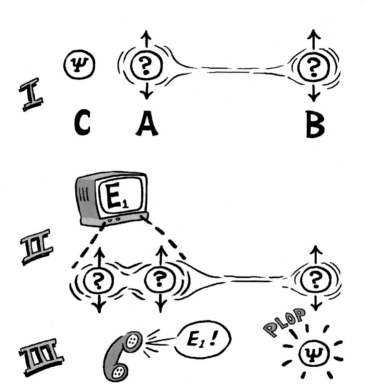

with either of them. Rather, it will be in the original state that A was in from the start! So simply by performing a measurement, Emmy has switched the role of B and C! Then teleportation has happened.

… to Tell Him Which Way to Rotate His Electron.

Actually, that is not quite the whole story yet. Emmy needs to write down the result of her measurement, and send it to Maxwell.

Depending on the precise way in which A and C turn out to be entangled, B might not be in quite the same state as C before, but rotated slightly. The result from Emmy's measurement tells Maxwell which way B has to be rotated in order to have precisely the original qubit of C.

Okay, we need to comment on what just happened: the quantum teleportation did not actually move the electrons from one place to the other. It is just that we have transferred all information of one electron to another one far away (from C to B). The teleportation process begins when Emmy makes the measurement of how A and C are entangled. It ends when she has told Maxwell the result of her measurement. She needs to do this with conventional methods, which is why teleportation does not happen faster than light!

There Is No Cloning with Teleportation!

It is also important to note that teleportation is not copying. Because Emmy performs a measurement on the electrons A and C, she changes their state. Before the measurement, C has some qubit, but after the measurement, C is in some strange entangled state with A, not in that qubit state any more. It is B, which carries the qubit after the teleportation process has finished. So in order to teleport a state, you necessarily need to destroy the original information. That might actually be a relief – if some day in the future we manage to actually teleport whole people with this method instead of just electrons, we should not have the problem of accidentally copying them, having several versions of the same person around.

By the way: although quantum teleportation sounds like science fiction, this kind of teleportation has actually been done. A successful teleportation of the spin of a single electron has been performed over several kilometers! Sounds like good news. But consider: in a human body there are about 10^{28} electrons, not to speak of all the protons and neutrons. So it is still a long way to go until we don't have to commute by car any more…

"At last! We have the technology to waste an infinite amount of time, in an infinite amount of parallel worlds, playing numerous games at once!"

Qubits
How to Build a Quantum Computer

One of the most interesting properties of electrons is their spin (↗[1]). It is like a little arrow which is attached to the electron, making it into a tiny magnet. The tip of the arrow is in the direction of where the North Pole of that magnet points to.

The Spin States: Up or Down?

If you measure whether the spin points into a certain direction – say, up- or downwards – then there are only two possible answers. These are "The spin points upwards" and "The spin points downwards". There is nothing in-between, these are the only two possible outcomes of the measurement. In particular, the result will never be "The spin points sideways". The British physicist Paul Dirac has invented a neat little notation for these two possibilities. One writes these as $|\uparrow\rangle$ for "upwards" and $|\downarrow\rangle$ for "downwards".

The Qubit: Up, down, or In-Between

Now, this being a quantum system, the spin of the electron does not necessarily have to be in either the state $|\uparrow\rangle$ or $|\downarrow\rangle$, it can also be in a superposition of the two. Just as Schrödinger's cat can be both alive and dead at the same time (↗[2]), so the electron's spin can be both up and down, with certain possibilities. In fact, the spin can be any linear combination of $|\uparrow\rangle$ and $|\downarrow\rangle$. One writes this as

$$|\psi\rangle = p|\uparrow\rangle + q|\downarrow\rangle$$

The numbers p and q mean the following: If an electron has a spin like $|\psi\rangle$ with some numbers p and q, then, whenever you measure the z-component of the spin, you will get the result "upwards" with a probability of p^2, and the result "downwards" with a probability of q^2. Well, since these are the only two possibilities, the probabilities better add up to 100%. Physicists don't like to use percentages, they rather like to use fractions between 0 (which is completely unlikely, i.e. 0%) and 1 (which is completely certain, i.e. 100%). So p and q need to be such that the sum of their squares adds up to 1. Every state of the electron's spin looks like some $|\psi\rangle$, maybe with different p and q – as long as p^2 and q^2 add up to 1!

If you have worked with computers before, you might have come to the following realization: If the spin can be up or down, then this seems perfect to store information. All data in our computers is expressed as a series of 0's and 1's – called bits. But whether you use 0's and 1's, or "on" and "off", or "current flows", "no current flows" does not matter – that is just the technical realization. The important point is that you need two different states, which encode the bit of information.

↗[1]: *"Spin" on page 187*
↗[2]: *"Schrödinger's Cat" on page 155*

That's what the spin of the electron can do – if you measure it, it can be either "up" or "down". So to store information, you only need to have lots of spins, which are either in the state $|\uparrow\rangle$ or $|\downarrow\rangle$. But remember: spins are not restricted to these two values, but can also attain any kind of superposition of these two values! This is why the spin of an electron is not called "bit", it is called "quantum bit", or qubit for short.

How to Play Angry Birds and Battlefield 3 at the Same Time

A computer which could store and manipulate qubits instead of normal bits is often called a quantum

ber five is represented with four bits as 0101, so in the quantum computer that would be $|\downarrow\rangle|\uparrow\rangle|\downarrow\rangle|\uparrow\rangle$. Similarly, the number 7 would be $|\downarrow\rangle|\uparrow\rangle|\uparrow\rangle|\uparrow\rangle$. But you could use any other qubit as well, for instance a qubit of the value

$$|\rightarrow\rangle = 1/\sqrt{2}\left(|\uparrow\rangle + |\downarrow\rangle\right)$$

The four qubits could have the value $|\downarrow\rangle|\uparrow\rangle|\rightarrow\rangle|\uparrow\rangle$ – then, the two numbers 5 and 7 would be stored, at the same time! If you were to read the memory now (that would be like measuring "all the spins"), you would get either 5 or 7, both with a 50-50 chance. But, and this is the important thing, performing calculations in a quantum computer is not mea-

$$|\rightarrow\rangle = \frac{1}{\sqrt{2}}\left(|\uparrow\rangle + |\downarrow\rangle\right) : \begin{matrix} 50\% \ |\uparrow\rangle \ up \\ 50\% \ |\downarrow\rangle \ down \end{matrix}$$

computer, and it works fundamentally differently from the ordinary machines that we have in our offices and at home.

Because a qubit can also have the values "up" and "down", a quantum computer can do everything a normal computer can – and more! For instance, it can store more than one number in a piece of memory.

For example, let us agree from now on that "spin up" means "1", and "spin down" means "0". Then, with a memory made out of four qubits, you could store, for instance, the number 5. In binary, the num-

suring. So you can perform several different computations simultaneously, without collapsing the wave function of the spin ($\nearrow^3$). By the way, this is a question for the computer geeks: which numbers would be stored in the memory if the qubits read $|\rightarrow\rangle|\rightarrow\rangle|\rightarrow\rangle|\rightarrow\rangle$?

A Danger for Online Safety?

To construct actual algorithms which make use of the power of quantum computers is actually quite complicated, but it can be done. Theoretically, there are several calculations which one could perform much much faster than with ordinary computers.

$\nearrow^3$: *"Wave-Particle Duality" on page 143*

This is quite important, for instance for online cryptography. The modern encryption algorithms are so safe, because they rely on the fact that apparently it is really, really hard to factorize a 90,000-digit number into its two prime factors. If you were to simply try all possibilities out by

factorization but on something else, we would be safe again. In fact, there are such algorithms, which quantum and normal computers would be equally bad at solving. We just don't use them right now on a global level, because it is not necessary.

Secondly, so far nobody has actually built a working, efficient quantum computer. There are some experimental designs, working with a few bits, but nothing ground-breaking (at least not at the time

$$|\downarrow\rangle \, |\uparrow\rangle \, |\downarrow\rangle \, |\uparrow\rangle \iff 0101 \iff 5$$

$$|\downarrow\rangle \, |\uparrow\rangle \, |\uparrow\rangle \, |\uparrow\rangle \iff 0111 \iff 7$$

$$|\downarrow\rangle \, |\uparrow\rangle \, |\rightarrow\rangle \, |\uparrow\rangle \iff \begin{array}{l} 50\% : 0101 \iff 5 \\ 50\% : 0111 \iff 7 \end{array}$$

brute force, a normal computer would need about a century for that task. A quantum computer would achieve this much, much faster – a great risk for all encryption!

But rest assured, your online banking data is safe from encryption breaking (as long as no backdoor has been programmed into the banking code…), for two reasons: Firstly, even if we had quantum computers, one could simply use another encryption algorithm. If that would not rely on the prime

of writing this book). For the real deal, one would have to use lots and lots of qubits, and not measure them. The power of qubits is that they can be in superposition of two bits, but only as long as nobody looks at them. It is really hard to shield them from the environment in such a way that they can be manipulated without measuring them, though. So in a manner of speaking, until we manage to keep Schrödinger's Cat reliably alive and dead for a long time, quantum computers are not something to reckon with yet.

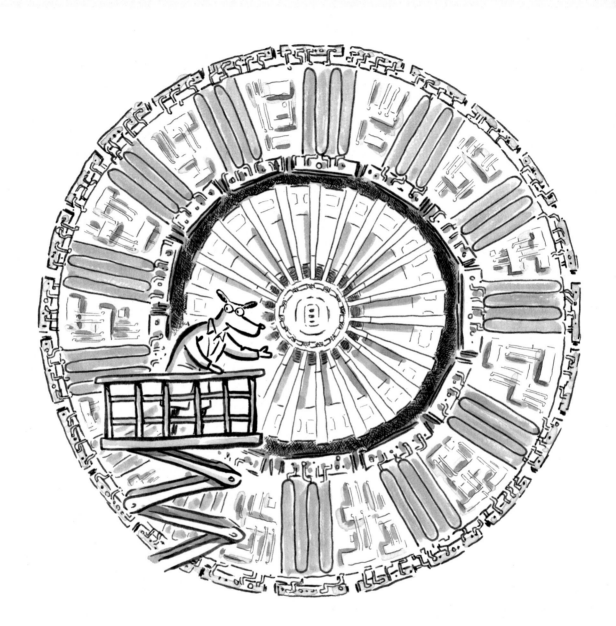

IV – Particle Physics

"When I was a small puppy, I loved smashing things together. It was kind of a hobby. Just imagine my surprise when I found out that that's what particle physicists are doing all day! In large particle accelerators, they bring the smallest particles up to incredible speed – and then they bash them together! The biggest accelerator is even called "Large Hadron Collider". It is located about 100 yards underground and has a circumference of about 17 miles – isn't that amazing? Basically that's how you find out what's matter made of: you find the smallest elementary particles by bashing larger particles together and checking if they break into something smaller.

Needless to say, crashing particles into each other to see what is in them is one of my favorite things today (besides building rockets)!

As it turns out, everything we know is made out of the same bunch of elementary particles. Quarks, electrons, photons, and the like – they are the basic building blocks of our universe. And a new one has just been discovered recently: the Higgs particle.

On the next few pages we will show you some of the most amazing particles, and what they do. I hope you'll be as fascinated by them as I was when I first heard about them."

"Delightful. But where does it stop?"

Atoms vs. Elementary Particles
Crack and Check, Crack and Check ...

Since the dawn of Humankind, we have tried to find out what the world – and everything in it – is made of. The idea was that there should be something fundamental – a set of elements. Everything was then supposed to be built out of these elements. For the ancient Greeks, the world was made of fire, water, earth and wind. Not too bad as a first guess. But today we know that you cannot mix fire, water, earth and wind to create a tasty pizza, for example.

of it will be chlorine and 39% will be sodium, as sketched in the first image.

John Dalton extended this discovery to the "law of multiple proportions": If you take two different substances, mix them and create two different new substances out of them, then the ratio of matter of the ingredients is always a ratio of integer numbers. Okay, even Emmy will be confused after reading this sentence for the first time. Let us look at an ex-

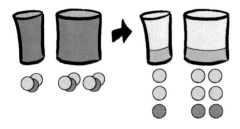

Law of definite proportions

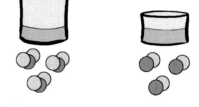

Law of multiple proportions

Today we have a somehow larger set of 118 elements. Each of elements has very distinct properties and a smallest unit – the atom. Let us briefly sketch the road to the discovery of atoms. The idea and the expression "atom" were formulated by the Greek philosopher Democritus in the 5th century BCE. The scientific hunt for the atoms, however, started in 1799 when Joseph-Lois Proust formulated the "law of definite proportions". It states that no matter how much of a substance you take, the mass ratio of its ingredients is always the same. For example, no matter how much salt you take, 61%

ample: You can mix hydrogen and carbon to either methane (CH_4) – the gas that cows burp – or wax (e.g. $C_{20}H_{42}$). If you take the same amount of carbon atoms (20), you need $4 \cdot 20$ hydrogen atoms and therefore the ratio of hydrogen atoms in methane compared to wax is 80:42, or 40:21. The image should help understanding the idea of that law.

Elements – Built out of Atoms

The fact that something shows up in discrete values is really a strong hint towards some smallest build-

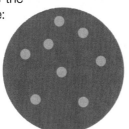

ing blocks of nature – the atoms. So we have a set of elements, each with unique properties, with no way to split it further by chemical reactions. The elements' smallest units, the atoms, were studied in detail by Dalton and his colleagues. Something interesting was observed by Dalton when checking the atomic masses: it always seemed do be an integer multiple of the mass of a hydrogen atom. It was the same effect once more: multiples of a certain unit. Could that mean that even atoms have an underlying structure and consist of something else?

It was J. J. Thomson who elicited a building block out of the atoms, which were yet believed to have no further constituents. By using cathode rays, he created beams of electrons. His experiment worked similar as the old TV screens (the non-flat ones) do. A wire is heated up by a current, the electrons are released from the wire and then accelerated by a voltage ($\nearrow$[1]). So there were tiny pieces within the atoms – electrons! Thomson though of a model of atoms which is used to be called "plum pudding model" – electrons are spread in a pudding of positive charge and form an atom.

But this model of an atom should not last long. In 1922, Ernest Rutherford replaced it with an impor-

THOMSON'S MODEL

tant update. He observed that the structure of atoms must be completely different. But how to observe what you cannot see?

Marble Games as Microscopes

There is a big problem with observing very small things: You simply cannot see them. Atoms have a size of about 10^{-10} m. Neither can you directly see such structures, nor can a microscope help you out. The Rutherford scattering experiment had a different approach. Imagine that you have an object that you cannot see because it is hidden. Then you let a bunch of marbles roll towards that object. The marbles will be scattered in different directions. These directions will depend of the shape of your hidden object. So by analyzing the distribution of scattered marbles you can get an idea of the structure on which you scattered. You can do the way Erwin would do it by calculating complex formulas or Maxwell-style by comparing the distribution to other distributions of reference objects with a known shape. Rutherford did not use marbles, but radiation consisting of α-particles ($\nearrow$[2]) which he was shooting on a thin foil of gold. The outcome of Rutherford's experiment is sketched in the image next to Thomsons model: The positive charge is concentrated in the center in the atomic nucleus, which carries almost the whole atomic mass, and is surrounded by the electrons. And in between, there is nothing.

RUTHERFORD'S MODEL

$\nearrow$[1]: "Particle Accelerators" on page 249
$\nearrow$[2]: "Alpha, Beta and Gamma Rays" on page 171

A new insight about the structure of the nucleus was given by James Chadwick's experiments in 1932. He found a particle with the same mass as the lightest atom, but without any charge: the neutron. The structure of atomic masses which was found 100 years earlier by Dalton could now be explained: atomic nuclei are built out of protons and neutrons. And these nuclei are surrounded by electrons. This is the same idea of atoms that we still have today.

Mini Marbles out of Super Guns

So, is this the end of the story? Obviously not as there is still a page remaining. The answer depends on the question: The story of atoms has ended. But the story of atoms is no longer the story of elementary particles. We know already that atoms are not fundamental. Instead, they are made of protons, neutrons and electrons. That is quite convenient. Instead of 118 elements you have three building blocks of our universe. But also this picture did not last for long. As physicists are like curious kids, they want to play with their marble microscope. They replaced the α-rays with electron rays. And they increased the energy with which they hit

the atomic nuclei. While Rutherford's scattering experiment could still be placed on a table, the electron gun and the detectors for the scattered electrons are slightly larger. The image shows an important experiment which was performed at the Stanford Linear Accelerator (SLAC, ↗1). As a result of several scattering measurements, it was found that protons themselves have an inner structure. The energy was high enough to break the protons! The results agreed with a model (Erwin style!) where the protons consist of an inner structure: quarks (↗3)!

You see that the game of breaking things and checking if there is something else inside has been played ever since. And as soon as "nothing else is inside" you have something that you can call an elementary particle. But as the past has shown: the status of what is actually elementary depends on how good your microscope is. This is the reason why people built giant particle accelerators to check if maybe there is even something inside quarks. Everything we know today as elementary, point-like particles without any extra inner structure is summarized in the chapter about the "Standard Model" (↗3). But: the search with accelerators will go on!

↗3: *"Standard Model of Elementary Particles" on page 213*
Image: US DOE

"I've got to eat more fiber. Lately, even neutrinos are having a hard time passing through me."

The Neutrino
So Light and so Hard to Catch

Neutrinos belong to the fundamental matter particles of our universe and are described in the Standard Model (↗1) together with many other particles. So why should neutrinos get their own chapter? Is there something that only they can do?

First of all, there are many things they can not do. As neutrinos carry neither electric charge nor color charge (only quarks do so), they interact neither via the electro-magnetic interaction, nor via the strong interaction. And as they are super light, they will barely act via gravity. The only thing that is left is the weak interaction. And the name says it: it is weak. Very weak. This means, neutrinos barely interact with anything. This is not too bad. Did you know that each second 1,000,000,000,000 neutrinos hit every cm^2 of our body? If they interacted like α, β or γ rays (↗2) we would be dead within seconds. But they simply pass through our bodies as if there was nothing. If a particle does not interact, it does also not lose its energy. Hence, it is hard to stop it. Really hard: you would need a lead wall one light year thick to stop 50% of the neutrinos passing through it.

Neutrino Discovery

If they are so hard to catch, how to we know that they exist? In the 1920s, physicists desperately searched an answer to a problem that drove them crazy: In radioactive beta decays, protons within an atomic nucleus convert into a neutron or vice verse. To conserve the charge, a positron or electron (β particle) is emitted. People observed these β decays and measured the energy of the β particles. In contrast to the energy of α rays, which always have a fixed value, β rays showed a broad distribution of energies. This means that the β particle did not carry the whole energy which should be released in these processes. Instead, it seemed to share it with another particle carrying the additional energy. But this could only be true in case this additional particle was invisible, would not interact and would carry no mass. But imagine somebody tells you that a huge elephant is standing in front of you, which you can neither see nor feel. This is hard to believe, isn't it? Every theory is only solid if it can be experimentally proven. Fortunately, the neutrino (the mystic invisible second particle) was interacting at least a little bit. That meant: building a detector to search for neutrinos would sooner or later (later in this case) lead to an observation. In these neutrino detectors, the β decays were happening in the reverse: a neutrino hitting a proton converted it into a neutron and a positron. It worked, and the neutrino was finally discovered in 1956.

So far, we have only been talking about one neutrino. But the Standard Model knows three: one

↗1: *"Standard Model of Elementary Particles" on page 213*
↗2: *"Alpha, Beta and Gamma Rays" on page 171*

for each type of charged Lepton (Electron, Muon and Tau). Each charged lepton will only interact with the corresponding neutrino. So for the β decays, only electron neutrinos were involved. But when for example a muon decays, a muon neutrino will pop out. There are also the antimatter partners of neutrinos: anti-neutrinos. This is to keep the matter/antimatter balance (↗³): When a positron (antimatter) is created in the β decay, a neutrino (matter) comes along. And if it is an electron, an anti-neutrino is created.

it's that many neutrinos hitting Earth and us. To estimate the number of neutrinos which should hit us, you need a good model of the sun and the things going on inside. Testing the model needs a measurement, so that's what physicists did: they built neutrino detectors and counted neutrinos. Remember the light-year thick block of lead to catch all these neutrinos? Well, physicists just tried their best and built detectors as large as possible. Some of these detectors work as the one from the neutrino discovery: reversing the neutrino production. Others use an effect called "Cerenkov Radiation".

If you ever got to know a nuclear reactor from the inside you might have seen this type of radiation. It is the blue shiny light in the water where the fuel rods are stored. During β decays inside the rods, electrons and positrons (we will say electron from now on, but it could also be a positron) are emitted. And they are fast. Very fast. Faster than the speed of light. "Ha", you might think, "nothing can be faster than the speed of light. I know that and Einstein knew that as well." This is true in case you are in the vacuum where light defines the high-

Neutrino Sources

Remember the gigantic amount of neutrinos flowing through all of us, that we talked about earlier? Have you wondered where they all come from? They are produced in the sun, the place where protons fuse to deuterium. During the fusion, a proton becomes a neutron and an electron-neutrino is spit out of the sun along with a positron. As the sun is quite big,

↗³: *"Antimatter" on page 217*
Image: Kirstie Hansen / IAEA

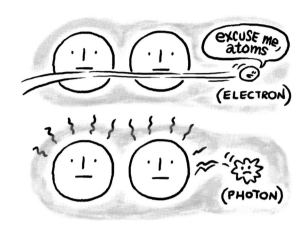

(ELECTRON)

(PHOTON)

when they transit from air to water or glass. So we have an electron traveling in water faster than light. Due to its electromagnetic interaction, the electron will "shake" the atoms of the water. This will make the atom's electrons oscillate, and oscillating charged particles create electromagnetic radiation: light. Usually, all atomic shakes cancel each other. But in case the electron shaking the atoms is faster than light, a cone of "Cerenkov light" will be emitted. This is similar to shock wave cones you can observe when an aircraft flys and breaks the sound barrier.

Now let us use Cerenkov light to detect neutrinos! All we need is water and light detectors. If a neutrino interacts within the water via the weak interaction, it will kick out an electron from a water molecule, so from a hydrogen or oxygen atom. This will then travel through the water with a very high speed and emit Cerenkov light which can then be detected. The direction of the Cerenkov light cone even tells us something about the incoming direction of the neutrino. To build a really good neutrino detector, you will need a lot of water. What about 50 million liters? To shield your detector against other sources of radiation (such as cosmic radiation, ↗4), it is best to build it underground, inside a mountain for example. A nice example for such a giant neutrino detector is SuperKamiokande, located unter Mount Kamioka in Japan. One of the main goals of the SuperKamiokande experiment is to measure a phenomenon called "neutrino oscillation" (↗5). And this oscillation is in fact something that only neutrinos can do. So this own chapter for the neutrinos is justified, don't you think so?

est speed ever possible. But in a medium such as water, things can be different. How is that possible?

How to Be Faster than Light

The maximal speed with which you can travel is determined by your interaction with the environment. Now let us determine the maximum speed of a baby which tries to crawl towards a Christmas tree. Suppose the baby is moving in an environment called family. They interact: every now and then someone will pick up the baby, hug it and put in back on the ground. This effectively slows down the baby. And the same thing happens to light in water: atoms absorb and re-emit it and effectively slow it down. The ratio of "speed of light in vacuum" to "speed of light in the medium" is called "refractive index" n. You might know it from optics: It is the number causing kinks in the light beam

↗4: *"Cosmic Radiation" on page 257*
↗5: *"Neutrino Oscillations" on page 261*

"Where does this fit in?"

Standard Model of Elementary Particles
So Far the Best Manual for Our Universe

A particle physicist cannot think about retirement before one big project is finished: the manual of our universe. It is supposed to contain all the building blocks of nature as well as a description of what they are doing. This manual is continuously being updated, still far away from being complete, but already very powerful. It is called the "Standard Model of Elementary Particle Physics". Technically, it is a renormalizable quantum field theory: the field quanta correspond to elementary particles. The fields and particles can be divided into two groups. Particles from the one group are the matter particles. As they all have a spin ($\nearrow$[1]) of 1/2, they belong to the group of "fermions", particles with half-integer spin. The other group of particles is somehow different. It also contains particles, called bosons due to their spin of 1 (there is also a boson with spin 0, the Higgs boson ($\nearrow$[2]), that plays a somewhat different role), but their purpose is different from the classical image of "sticking them together like LEGO bricks to get something big". These bosons are also called "force carriers" as they are exchanged between the fermions and mediate inter-

actions. Let us try to illustrated that image with an admittedly simplified example.

Interactions Caused by Particle Exchange

Imagine Erwin and Maxwell standing still on skateboards. Maxwell throws a banana towards Erwin. The momentum conservation ($\nearrow$[3]) makes him move in the opposite direction of the thrown banana. Once Erwin catches the banana, he will take over the banana's momentum and also move in its former direction. They can keep playing the game back and forth. Now imagine Maxwell and Erwin are the fermions, and the banana is a boson, which mediates a repulsive force. For attractive forces you need a somehow more advanced example like a thrown boomerang.

In total there are three fundamental interactions incorporated by the Standard Model: the electromagnetic interaction, the strong interaction ($\nearrow$[4]) and the weak interaction ($\nearrow$[5]). Feel free to replace the word "interaction" with "force" to get the idea. But if you take a closer look at the interactions you will see that they are more than just making things

---

$\nearrow$[1]: *"Spin" on page 187*
$\nearrow$[2]: *"The Higgs Mechanism" on page 241*
$\nearrow$[3]: *"Conservation Laws" on page 51*
$\nearrow$[4]: *"The Strong Interaction" on page 229*
$\nearrow$[5]: *"The Weak Interaction" on page 233*

attractive and repulsive. By the way: do you miss something? Gravity? You are right, and this is one of the weak points of the Standard Model: it does not explain gravity (for more information, see the chapters about "String Theory" (↗6) and "Quantum Gravity" (↗7)). There are still unsolved problems to get gravity in the same mathematical framework as the other interactions. But at the level of elementary particles we can safely ignore gravity: it's impact is simply far too small.

While the electromagnetic interaction is responsible for almost everything we experience in daily life (cell phone calls, microwave ovens, heartbeats, muscle movements, visible light, …) the strong and weak interaction are more hidden. The strong interaction does the job of holding atomic nuclei together for which we should thank it. Otherwise we would evaporate immediately. The weak interaction allows many nuclear reactions such as nuclear fusion (↗8), the heating mechanism of our sun. So: no weak interaction, no sunlight, no life on Earth. Good to have it!

Not every fermion can take part in every interaction. Each interaction has a dedicated "charge" that the fermion has to carry. For the electromagnetic interaction it is the classical well known electric charge. We know that particles with positive charge attract particles with negative charge, repel those which are positive as well and do nothing with neutral ones. The strong interaction acts on a charge called "color charge" (↗4). Fermions carrying this charge are called "quarks". The fermions which are not strongly charged are called "leptons". The charge of the weak interaction (the weak charge) is carried by all fermions of the Standard Model. This means every particle can interact weakly. For those fermions which are neither quarks nor charged electrically, this is the only way to interact at all. We call these particles neutrinos.

The collection of Standard Model bosons consists of photons, W and Z bosons, gluons and the Higgs boson. The massless and electrically neutral photons are the mediators of the electromagnetic interaction. As they are massless, the range of the electromagnetic interaction is – even though its strength drops with distance – infinite. Think about Erwin, Maxwell, the bananas and the skateboards. Massless bananas can be thrown quite far. Things are different for the weak interaction: its mediator, the W and Z bosons, are very heavy. Look at the picture with Erwin, Maxwell and a melon. It cannot be thrown that far, right? The masses of the W and Z bosons cause the weak interaction to be of a very short range. For the strong interaction and the mediated gluons, things are different again. Gluons are massless, but the strong interaction still has a very short range only. The reason is that gluons themselves carry a color charge. You can read the chapter about the strong interaction to get to know

↗6: "String Theory" on page 303
↗7: "Quantum Gravity" on page 279
↗8: "Nuclear Fusion" on page 175

the reason how this leads to a short range. The last missing boson, which you might already know, is the Higgs boson. It still is a boson, but in contrast to his colleagues it has a spin of 0, not 1. It is also not a mediator of an interaction, but it is responsible for giving all particles of the Standard Model their masses.

A Whole Zoo of Particles

On this page you can see an overview of all particles of the Standard Model: The quarks, the charged leptons and the neutral leptons (neutrinos) are arranged in three generations. In the first generation you find all the particles that you need to describe life on Earth: You can build protons out of two up quarks and one down quark. For neutrons you need two down quarks and one up quark. With the protons and neutrons you can build all atomic nuclei and then you just have to add the charged lepton of the first generation, the electron. The other generations have the same properties as the first one, but their particles have higher masses. This allows them to decay (↗9) into the lighter ones. In history, not all particles have been known from the very beginning. Sometimes the experimental physicists have found a new particle and the theoretical physicists had to add a generation. And sometimes the theoretical physicists had a good reason to propose a particle (such as the top quark as partner of the bottom quark) which has then found later on by an experiment.

Symmetries as those between the generations have always been a fundamental property of physics. They also allowed to describe several interactions with the same formalism and motivate physicists to find one theory to describe everything, maybe even gravity, with the same formalisms. This is the business of theories like string theory (↗6), supersymmetry (↗10) or Quantum Gravity, for instance (↗7). They could also help to find answers to questions that the Standard Model is not yet able to solve: Why are there so many particles? What is Dark Matter (↗11)? And how does gravity work?

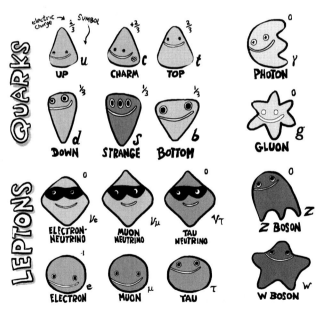

↗9: "Particle Decays" on page 221
↗10: "Supersymmetry" on page 299
↗11: "Dark Matter" on page 133

"My whole life I've been looking for someone like you --- what are the chances I should meet your antimatter double?"

Antimatter
More Science than Fiction

Antimatter sounds fancy, futuristic and dangerous to many people. Just think of the movie "Angels and Demons" with Tom Hanks, where a freak steals antimatter from a lab to build a giant bomb. The movie contains both truth and fiction concerning antimatter – but what is it actually?

It is defined in a way that sounds… well, technical. "Antimatter is like normal matter but with the signs of all quantum numbers inverted". We can stick to the most important quantum number: the electric charge. Let us start with an electron. It has a negative charge. And now invert its charge. Ta-da, you have a positive anti-electron, called "positron". It will have the same mass as an electron, but its charge is just positive.

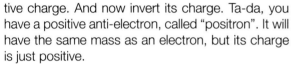

How to Get Antimatter?

Getting antimatter is not an easy thing. You need to make use of Albert Einstein's famous $E=mc^2$. This formula says that if you need something massive – such as antimatter – just take some energy and convert it. The more massive the things are that you want to create, the more energy you need. But here comes another conservation law ($\nearrow^1$) of nature: Antimatter does not simply come out of energy. At least not alone. It comes along with a corresponding amount of matter.

Heavy anti-particles need a lot of energy. People build big machines to pump energy into particles, smashing them together and let them release their energy into matter/antimatter pairs. These machines are called "particle accelerators" ($\nearrow^2$). If you just want to make light antimatter, such as an electron/positron pair, a high-energetic photon is sufficient. You can get it from a particular type of ionizing radiation, the γ-radiation ($\nearrow^3$). This radiation is released for certain types of nuclear decays ($\nearrow^4$). Other nuclear decays directly produce positrons: The β-radiation ($\nearrow^3$). You can impress your friends with your personal source of positron radiation. Just visit the supermarket and buy a banana and you are done. Why is that? Bananas contain potassium. There is a small fraction of potassium atoms which carry one more neutron in their nuclei than others. And part of these decay via positron radiation. No reason to worry: your stomach is full long before you have eaten an amount of positron radiating bananas which can harm you.

If there is antimatter all around us (bananas are not the only source) the question is: what happens to it? It disappears. Not into the void, but once it meets a corresponding matter partner, it will annihilate. Annihilation means: All the mass goes back into energy. In most of the cases: two photons. This is just $E=mc^2$ backwards. In case of a positron, this will quickly happen as there are many electrons in the matter surrounding us.

You can make use of such annihilation processes in medical applications ($\nearrow^5$). If you have too many of

$\nearrow^1$: *"Conservation Laws" on page 51*
$\nearrow^2$: *"Particle Accelerators" on page 249*
$\nearrow^3$: *"Alpha, Beta and Gamma Rays" on page 171*
$\nearrow^4$: *"Radioactive Decay" on page 167*
$\nearrow^5$: *"Radiation Therapy" on page 265*

these annihilation processes, they'll release a lot of energy. And here we go: the antimatter bomb from "Angels and Demons".

Antimatter Bombs – Are They Real?

What happens if you take 0.5 grams of antimatter and bring it together with 0.5 grams of normal matter? Just insert 2 times 0.5 grams (matter plus antimatter) into $E=mc^2$ (which means multiplying the mass with the squared speed of light c to get the energy) and you'll get 90,000,000,000,000 Joules of energy. Is that a lot? You can convert it into a more common unit: the sausage. Eating one sausage adds 300 calories (actually, what we use to call "calorie" is in fact a kilocalorie, so 1000 calories) of energy to your body. This corresponds to 1,200,000 Joules. So if you want to add the same amount of energy that is released in the annihilation process of 0.5 grams of antimat-

ter to your body by eating sausages, you'll need to eat 72 millions. Impressive. Okay, while a sausage might be a common unit, eating 72 millions of them is not a common process. You can also compare the amount of energy released in the annihilation of 0.5 grams of antimatter to the one released in the explosion of the nuclear bomb "fat man" which detonated on the 9th of August 1945 over the Jap-

anese City Nagasaki. You see: Already 0.5 grams of antimatter can cause a lot of trouble. Fortunately, it is technically nearly impossible to produce and store so much antimatter.

Storing a single antiatom is already quite a big deal. The reason is that you cannot simply put antimatter in a box and store it. Your box – at least if you buy it in an ordinary matter shop – and the antimatter will annihilate. What you can do: Find an empty space or make one, called vacuum (↗6). Produce your antimatter right in there. It won't have surrounding matter for annihilation. So far, so good. But most likely, it will move. It will move out of the vacuum. And you know what happens then. To make antimatter stay inside the evacuated area you cannot drag it with your hand, but with electric and magnetic fields.

So antimatter is difficult to store. Creating ordinary anti-particles is not that of a big deal. But combining these to anti-atoms, or even building anti-molecules out of those, it gets really tough. You need a radioactive source providing positrons. You would also need a particle accelerator to provide you with antiprotons. Both you will get separately. But you need to slow your antiprotons and positrons down, and then make them approaching each other slowly. If you are lucky, they will combine into

↗6: *"Vacuum and Air Pressure" on page 23*

anti-hydrogen. This is the most simple anti-atom you can imagine. Several experiments (↗7) produce, store and analyze antiatoms.

Studying the Antiworld

But what makes these guys so interesting? We said that antimatter is the same as ordinary matter, but with opposite charge. The rest is the same. Well, that's what physicists assume. But will antiatoms emit light with the same color as ordinary matter? Will a block of antimatter also fall downwards, attracted by gravity? We assume: yes. But we don't know. It could also be repelled and "fall upwards". There are many ideas for antimatter experiments, and physicists are just beginning to understand it. Many mysteries concerning antimatter are still unsolved. For example: why do we exist?

According to the big bang theory (not the TV series, the real theory!) the universe emerged from a single point containing an enormous amount of energy. This energy was converted into matter/antimatter pairs. And these again annihilated. The process went on and on while the universe was expanding. During the expansion, the universe and its content cooled down. Cool means: less energy. So the pair creation process stopped. If the universe started from energy only, and creation and annihilation work as we expect, the amount of matter and antimatter should always be equal. Is this true?

Well, obviously, if we take a look at the sky we see only matter. And a lot of nothing. But no antimatter. If there was antimatter outside in the universe,

even far away, we would see it colliding with ordinary matter from time to time. But we don't. And we don't know why. It seems as there is a little asymmetry in the matter/antimatter produced. And only due to this very small asymmetry (1 in 1,000,000 cases) we exist.

Oh, by the way: How can you distinguish between electrons and positrons? Physicists use magnetic fields to do the trick. Charged particles get deflected by magnetic fields. Depending on the sign of the charge, the deflection goes into the one or the other direction. This way positrons were observed for the first time in the cosmic radiation (↗8), a type of radiation hitting the Earth from far above. A particle was observed (how physicists observe particles which are too small to be seen is shown in the chapter about particle detectors, ↗9) that had all the properties of an electron, but the deflection in the magnetic field went into the "wrong direction". So: "wrong" charge. And there it was, the positron.

For each of the matter particles of the Standard Model (↗10), the theory that contains all the particles and the way they interact, a corresponding antiparticle exists. There is one remaining question: In case you meet a dog on the streets, how do you know that it is made of matter or antimatter? Putting the dog in a magnetic field won't help. The dog consists of atoms which are neutral. So: No deflection. Okay, what you can do is: you can argue that a dog on the streets that does not annihilate with the street is an ordinary dog, not an anti-dog. But why are we made of matter and not antimatter? This is pure definition. Nobody wants to call himself anti-something. That's why "anti" refers to the "other world".

↗7: http://home.web.cern.ch/topics/antimatter
↗8: *"Cosmic Radiation" on page 257*
↗9: *"Particle Detectors" on page 253*

↗10: *"Standard Model of Elementary Particles" on page 213*

Particle Decays
The Particles' Short Lives and Interesting Heritages

When we talk about elementary particles as the building blocks of nature, we have a picture in mind in which we stick all the particles together to build whatever we want. This is more or less true for up and down quarks (which form protons and neutrons) and for electrons. But what about all the other particles of the Standard Model (↗[1])? They will not leave you enough time to build something like a castle or an airplane. Before you can stick them together, they will decay. This means that they only exist for a very short time. Depending on the particle it can be as short as 882 seconds (in case of a free neutron) or 0.00000000000000000000000001 seconds (in case of a top quark). This lifetime is only defined on a statistical basis. If you take a bunch of neutrons and wait for 882 seconds until the end of their lifetime, the fraction of $1/e=0.368$, so about one third of them, will still be there. The rest decayed. The lifetime only gives you an average about the whole bunch, not a single one. It might decay immediately, but it might also live much longer than its colleagues.

Why Particles Decay

You might wonder why particles decay at all. The answer is quite simple: because they can. Unless there is any law of nature that forbids a process, it will happen. Such a law of nature is often the conservation of energy (↗[2]). The piggybank from the chapter about particle accelerators (↗[3]) will not decay until you add energy (via a hammer, for example) to break the molecular bindings that pre-served its structure. The broken pieces of the piggybank is what probably most people have in mind when they think about a piggybank decay. Such a classical picture is indeed true for things like decaying atomic nuclei. The nuclei of uranium atoms can for example simply break apart into two lighter parts. This process is known as nuclear fission (↗[4]).

Atomic nuclei are composite objects made of protons and neutrons (which are also objects composed of quarks and gluons). But real elementary particles like a top quark…are elementary. Into what should they break apart if there is nothing that they are made of? This might look like a problem. But if you ask a particle physicist about particle decay, he will probably tell you that this decay actually refers to a reaction or transition. Let us take a look at the decay products of a top quark: a b-quark and a W-boson.

The W-boson is the mediating particle of the weak interaction (↗[5]). And this weak interaction is the only interaction that can really transform particles into others. So instead of saying that the top quark decayed into a b-quark and a W-boson you can also say that it transformed into a b-quark via the weak interaction.

↗[1]: *"Standard Model of Elementary Particles"* on page 213
↗[2]: *"Conservation Laws"* on page 51
↗[3]: *"Particle Accelerators"* on page 249
↗[4]: *"Radioactive Decay"* on page 167
↗[5]: *"The Weak Interaction"* on page 233

Three conditions must be fulfilled for something like this to happen. First, the particle needs to carry a certain charge, the weak charge in this case. We also say that the top quark couples to the W boson. The second requirement is that the decay products – or more precisely, the sum of their masses – needs to be lighter. The differences between the mother particle's mass and the masses of the decay products are put into the system as kinetic energy (the energy of movement) of the decay products. And as a last condition, all quantum numbers have to be conserved (see chapter "Conservation Laws", ↗2).

These conditions allow free neutrons to decay, but not protons. In the process of the transition of a proton into a neutron (↗5) an up quark transforms into a down quark via the weak interaction. A down quark transforming into an up quark makes a neutron becoming a proton. Down quarks are slightly heavier than up quarks, making neutrons (consisting of two down quarks and one up quark) slightly heavier than protons (two up quarks and one down quark). Hence, a free neutron can decay into a proton, but not the other way around. "Wait!" you might think now, "what about the beta decay of nuclei? There, protons also transform into neutrons. How can that be?" It is true that within an atomic nucleus protons can decay. That is not easy to understand, but let us try: If a proton sitting inside a nucleus converts into a neutron, there is less repelling positive charge in the nucleus. The strong interaction always has to struggle with the electromagnetic interaction to keep the repelling protons together. In case the proton became a neutron, less binding energy is needed. And this energy can then, in the spirit of $E=mc^2$, be used to allow a proton to transform into a neutron. Phew. That was tough. If you want to learn

more about the way that particle physicists draw and calculate decay processes you should read the chapter about Feynman diagrams (↗6).

What to Learn from Particle Decays

The fact that particles decay can in fact tell you a lot about their properties. First, you know that lighter particles must exist. And second, by understanding the decay processes you learn a lot about the way that particles interact with others. Most of the time there is not only a single way that a particle can decay, but several. All you need is a lighter particle, an interaction that connects the two and the consideration of all conversation laws (↗2). The more possibilities a particle has, the shorter its lifetime is. In case there is no way to decay, it lives forever. The lifetime is further depending on the strength of the interaction that mediates it. The reason why a muon lives so surprisingly long is because it can only decay via the weak interaction, but not the strong

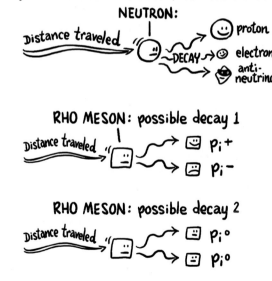

↗6: *"Feynman Diagrams" on page 225*

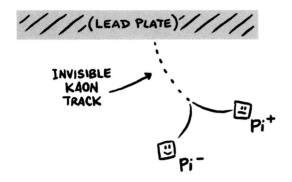

or the electromagnetic interaction. Each specific way of a particle decay is called "decay channel". Sometimes particles have quite different lifetimes even though they have the same decay channels. When physicists calculate lifetimes they have to consider an important aspect apart from the interaction strengths and number of decay channels: is there enough space for the particle to decay? It is not space in a sense of "enough room", but rather "enough possibilities to spread the energy which is released in the process." This space is called phase space. The larger the difference between the masses of mother and daughter particles, the more energy can be distributed. This is part of the explanation of the large difference of lifetimes between the neutron and the top quark: While the neutron's mass is only 1 MeV larger than the proton's mass, the top quark mass is about 89,000 MeV larger than the masses of the b-quark and the W-Boson. The Particle Data Group has listed all possible decay channels and lifetimes of elementary particles. You can check the impressive listings on their website (↗7) if you are interested.

A decaying particle might sound rather sad. But in fact if offers you a nice way to discover things! As described in the chapter of particle detectors (↗8) particles can only be detected by their interaction with a detector. Some detectors are only sensitive to charged particles and will hence not observe neutral ones. For example, cloud chambers make charged particles visible as tracks in the cloud. With the help of such a cloud chamber, the physicists Butler and Rocherster discovered the neutral kaon in 1947. The kaon is a particle made of a strange and an anti-down quark. As it is neutral, it did not leave a track. In fact, the kaon left no trace at all. But then, after traveling a few millimeters through the lead, it decayed.

The decay products, two charged pions, could be observed and measured. If you know the directions, the masses and the speed of the decaying particles, you can construct the so-called "invariant mass" of the two decay products, which is also the mass of the mother particle that decayed. In this case, it corresponded to the mass of the kaon. For many other particles, such a reconstruction via the decay products is the only way to observe them, as their lifetime is simply too short to be seen directly.

By the way: do you still think that the word "decay" should rather belong to a rotting zombie than a nice and friendly elementary particle? Then you are not alone. A group of physicists from the particle physics research facility CERN made a zombie movie. It is available for free download (↗9). Now, guess its name: "Decay".

↗7: http://pdglive.lbl.gov
↗8: *"Particle Detectors" on page 253*
↗9: http://www.decayfilm.com

*"Could they not have given us better directions
to the Feynman Lab?"*

Feynman Diagrams
Particle Skribblings with a Serious Meaning

From time to time, when a prehistoric cave is found, scientists discover little paintings and sketches on the walls. We can then get an idea of how people lived in these prehistoric days: for instance, one could deduce which animals were they hunting. If a few thousand years in the future people find the ruins of a lab with some scribbles on the wall, they will have difficulty figuring out if these were some silly paintings made by a little kid or a description of a complex process of quantum chromodynamics. The confusion can occur when the process of quantum chromodynamics is written in the language of so-called "Feynman-Diagrams".

diagrams allow to write these very complex equations and formulas down in the form of a few lines. All you need to do is to use lines for particles and stick them together according to rules of the Standard Model and the conservation laws of nature.

These drawings pose a problem, of course. A physical process happens in four dimensions: three for the position in space and one for the time. But if you draw something on paper, you only have two dimensions. You can go ahead and try drawing a three-dimensional cube on paper. And now a four-dimensional one. Not easy, right? Physicists decided to simplify the three dimensions used for space to one. The second remaining dimension on paper is then used for the time. All right, now we have spacetime. Let us put some particles in it!

The Language of Particle Physicists

It is (next to English) the main language in which particle physicists communicate. They use it to describe what particle physics is all about: the interactions between elementary particles. The framework describing the interactions between matter particles (fermions) via the exchange of force carrier particles (bosons) is the Standard Model (↗[1]). It can be used to calculate all different kinds of processes via formulas which can get quite complex. Feynman

Giving Lines a Meaning

All particles are indicated by lines: solid lines for fermions, wiggly lines for photons, curly lines for gluons and dashed lines for W, Z and Higgs Bosons. The fermion lines get an arrow in addition. It indi-

↗[1]: *"Standard Model of Elementary Particles" on page 213*

a subatomic landscape

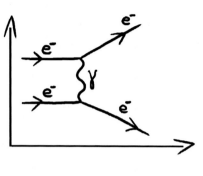

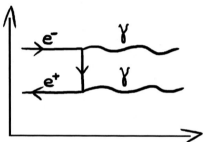

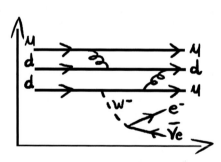

cates the direction in time that particles are

traveling. "Hey", you might say, "of course every particle travels forward in time!". That is true. But antiparticles are indicated by arrows that point backwards in time. It might seem weird. But mathematically there is no difference between antimatter propagating forward in time, and matter propagating backward in time.

You can now stick together bosons and fermions. A connection between particles is called "vertex" and is indicated by a dot. Do not forget to stick only together what is allowed to interact! For example: only charged particles can take part in the electromagnetic interaction and hence be connected to photons. And remember not to break any conservation law of nature (↗2)! Before we take a look at some examples we should talk about the concept of "real" and "virtual" particles. All the stuff that we can observe in nature, that we can "see" either directly or indirectly, is real. Real particles enter a Feynman diagram at

the beginning of the time axis and leave it at the end. "Virtual" is whatever starts in between and stays in between. If you describe a process of two electrons repelling each other as described by the electromagnetic interaction, the exchanged photon will be virtual and you will never get to observe it. Virtual particles play an essential role in the context of Heisenberg's Uncertainty Principle (↗3). Now let us finally take a look at three example processes. For each of them, the horizontal x-axis indicates time and the vertical y-axis space. This means events start on the left side and evolve towards the right.

The first example shows two electrons (e). They exchange a virtual photon and repel each other. You can see the repulsion as the two electron lines move apart in the vertical (space) direction after the photon exchange. A more complex process is shown in the second example: an electron and a positron (with the arrow pointing backwards) meet and annihilate (↗4) to two photons. This time, the photons are real as they leave the diagram on the right side. You can also see that at the end both the electron and positron are gone. The matter/antimatter conservation, which states

↗2: "Conservation Laws" on page 51
↗3: "Heisenberg Uncertainty" on page 151
↗4: "Antimatter" on page 217

that the difference in numbers of particles and antiparticles does not change within a reaction, is also nicely illustrated by the fermion line on the left: The line enters the diagram as electron and leaves it as positron. So the electron and the positron that meet and annihilate look like an electron which just went backwards in time after emitting two photons. The last example shows the process of a β-decay (↗5): a neutron is converted into a proton while a down quark from the neutron changes into an up quark by emitting a virtual W- boson which then decays into an electron (e) and an electron-antineutrino (v). Details about this process are explained in the chapter about the weak interaction (↗6). As you can imagine, there are many, more complicated processes that can occur. But we'll leave it at that.

More Than Just Lines!

Be aware that each line and vertex of a Feynman diagram corresponds to a mathematical expression which is used to calculate the properties of the corresponding process. For example, you can calculate the lifetime of a neutron from the third example. In general, Feynman diagrams provide the probability of these processes. You can take a look at (↗7) to see the complexity of the mathematical expressions which are derived from simple Feynman diagrams.

Let us finish with a very special type of Feynman diagrams which is shown in the last figure. Let us not talk about the process itself (you guess: not an easy thing) but rather about the story behind it. It starts in a pub close to the research facility CERN near Geneva, Switzerland. The theoretical physicist John El-

lis played a game of darts against a colleague. They made a bet that in case Ellis lost he had to put the word "penguin" into his next scientific publication. For a man working in the field of elementary particle physics, this is not an easy thing. His colleague left before the end of the game, but another of his colleagues took over and beat Ellis. He had no idea how to put the penguin into his paper until one evening he visited some friends and had an inspiring evening. And all of a sudden, he had an inspiration:

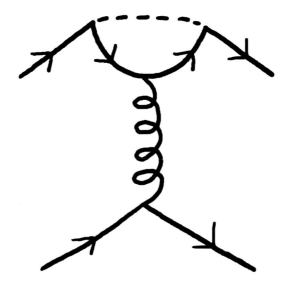

one of the Feynman Diagrams of his paper looked like a penguin! At least in his opinion. See it for yourself!

Can you see then penguin? You can try drawing it on top of the diagram, if you like. The whole story about the origin of the name "penguin diagram" can be found at (↗8).

↗5: *"Alpha, Beta and Gamma Rays" on page 171*
↗6: *"The Weak Interaction" on page 233*

↗7: http://en.wikipedia.org/wiki/Feynman_diagram
↗8: http://en.wikipedia.org/wiki/Penguin_diagram

THANK GOODNESS FOR THE STRONG FORCE
without it there would be...
no galaxies...
no stars...
no carbon...
no oxygen...
no complex chemical reactions...
no life.
no US.
no kidding.
STRONG FORCE
u
u
d

The Strong Interaction
Keeping Our Atoms Stable

Almost every interaction, every force that we experience in daily life is either an electromagnetic interaction, or a gravitational one. But there is one interaction which is much stronger than anything of electromagnetic or gravitational origin. You might think of something that superman uses to defeat his enemies. But this force is on the one hand immense and on the other hand quite modest. We do never observe it acting. But if it did not exist, we would all disintegrate immediately. It keeps our atoms together. But why do you need a very special force for that? Why would we all disintegrate without it? Let us have a look at how that works:

An atomic nucleus is made of protons and neutrons. We usually picture them with yellow/white and red balls. The yellow ones represent the electrically neutral neutrons while the positively charged protons are represented by the red ones. Protons and neutrons are so light that their interactions within a nucleus have not much to do with gravitation. But the situation is very different if we look at the electromagnetic force! While the neutrons do not care, protons do. Two positive charges repel each other. The Coulomb force, the electrostatic part of the electromagnetic force, is responsible for that.

And can we neglect it, as we did in case of gravitation? No, not at all. Atomic nuclei are very tiny, only about 10 fm (femtometer, 10^{-15} m). This implies a very short distance between two protons. And as the Coulomb force is proportional to the inverse of the squared distance, that makes it really strong. Here, really strong means about 2.3 N (Newton). To compare this to daily life: It is the force that you would need to lift about four chocolate bars with about two ounces of weight. "Not much!", you might say now. But remember that you are quite a lot bigger than a proton. And compared to the attractive gravitational force between the protons, the repulsive Coulomb force is a factor 10^{36} bigger. But as we all know, atomic nuclei are stable (except the ones decaying radioactively, ↗[1]). So there must be something which compensates the repulsion of the protons and keeps them bound together. This is the strong force.

The Colorful Will Be Strong

As only charged particles are affected by the electromagnetic interaction, we could ask ourselves which property a particle has to have in order to participate in the strong interaction. It keeps protons and neutrons sticking together, but does it also glue electrons to protons and neutrons?

↗[1]: *"Radioactive Decay" on page 167*

The property that particles need to take part in the strong interaction is called "color charge". It is a charge in a sense on "allowing to take part in an interaction", such as the electric charge is. But it has nothing to do with color, literally. We will soon see why the name "color" was established. The group of particles carrying a color charge is called "quarks" (↗²). Fortunately, electrons are no quarks. Otherwise, all the atom's electrons would be pulled inside the atomic nuclei. This means that we would immediately shrink to tiny balls, not able to do anything. This is as inconvenient as disintegrating.

No Distance Is Too Far

The strong force has a property that makes it really odd. It leads to a behavior that is completely contrary to our classical understanding of a force. Think about gravitation and electromagnetism: while a

bear falling from a tree gets dragged to the ground, a bear up in space does not. And while two magnets being really close can attract or repel each other quite strongly, two magnets far apart seem to be not interested in each other at all. The reason of both the bear's and the magnet's effect lies in the fact that the electromagnetic and the gravitational force decrease with the distance squared.

In contrast to that, the strong force remains constant over distance. This is not easy to explain. It has to do with the fact that the particles which are exchanged during the strong interaction, the gluons, do not only interact with color charged quarks, but also with other gluons. This is not the case for the photons, the exchange particles for the electromagnetic interaction (↗²). The fact that the strong force remains constant and therefore its range is infinite immediately raises a question: why does it only

↗²: *"Standard Model of Elementary Particles" on page 213*

keep protons and neutrons of a single atomic nucleus sticking together? Why are not all neighboring nuclei glued to one single superatom?

Creating Quarks from Field Energy

Let us check what would happen in case we try to separate two quarks that are glued – now you see where the name comes from – via a gluon. A constant force leads to a potential energy that increases linearly (twice the distance, twice the energy). So the further you drag them apart, the more energy is stored in the gluon. If you consider Einstein's $E = mc^2$, you will sooner or later reach a distance where you have enough energy in the gluon to convert it into the masses of a quark and an antiquark: $E_{gluon} = (m_{quark} + m_{antiquark}) c^2$.

And what can happen will happen: The gluon connection breaks and you will end up with two quark pairs with short distance instead of one with a large distance. This is the reason why you will also never be able to separate a quark from its friends. While it is relatively easy to kick an electron, bound by the electromagnetic force, out of an atom, you will immediately produce lots of new matter and antimatter if you try the same with a quark from a proton or neutron. Particle accelerators (↗3) such as the Large Hadron Collider make use of that to produce a whole variety of

quarks and antiquarks. The properties of the strong interaction explain why quarks do not exist as free particles, but will always be bound to at least another one. Physicists call this phenomenon "confinement". They also observed that particles built out of quarks (they call them "hadrons") can only exist as "white objects". What does that mean? Remember that quarks carry a color charge. While an electric charge can be positive or negative, a color charge can be either red, blue or green. These colors (again: please do not take it literally!) were chosen for a good reason. You might recognize them as the primary colors. Adding them leads to white. This means, a white hadron can be built out of three quarks with different colors. Such object are called "baryons". Another, more sneaky way to get white is to combine a color with it's anti-color. This is what antiquarks carry. Hence, also a quark and antiquark can form a white hadron, in this case called "meson".

If you ask yourself: why does the strong interaction have all that weird properties? We don't know. But if it had not, we would all not exist.

↗3: "Particle Accelerators" on page 249

The Weak Interaction
Weak but with Unique Power

What would you consider as an interaction? An interaction is, whatever causes a change. Interactions make you see, hear, feel, walk and live. They can produce heat or use energy to make machines work. In some cases, interactions seem to cause changes in objects. For example, they can make a tree grow. But if you look close enough, these interactions never change an object itself. They only change configurations: atoms from the air and the soil get rearranged to a tree. Every change that we observe in daily life is caused by a change of configuration, never by a change of an object on the level of an elementary particle. These changes are caused by attractive and repulsive forces and we can attribute them to either the strong, the electromagnetic or the gravitational interaction.

Neutron   Proton   W-Boson   Proton Electron   Anti Neutrino

The Only Interaction That Changes Particle Types

But what about an atomic nucleus that performs a β⁻-decay (↗1)? It emits an electron and an anti-neutrino, and at the same time converts one of its neutrons into a proton. Right, protons and neutrons are not el-

P-Tooo!

ementary, so you can still claim that there is simply a re-arrangement within the neutron going on. What actually happens is that one of the neutrons' down quarks gets converted into an up quark. But as the quarks, the electron and the anti-neutrino are elementary, something beyond a re-arrangement must have happened. The same quark conversion happens during nuclear fusion (↗2) in the sun, for instance. And this is exactly where we need a completely new type of interaction: the weak interaction. Let us take a closer look at the conversion of a neutron into a proton.

We see that after the down quark became an up quark, it emitted a W-boson. The upper sketch is more symbolic and the lower one is speaking the language of Feynman Diagrams (↗3). Each fundamental interaction is represented by the exchange of a boson (↗4) and the W-boson is, next to the Z-boson, the exchange particle of the weak interaction. The W is unique in two aspects: it is charged. And as charge is conserved in physics (↗5), the W-boson must take it away from another particle when it is exchanged (and also bring it to another one). The particles in a process of the weak interaction via a W-boson exchange have to change their electric charge. But as the charge is part of the definition of an elementary particle, a weak interaction by the exchange of a W-boson always comes

↗1: *"Alpha, Beta and Gamma Rays" on page 171*
↗2: *"Nuclear Fusion" on page 175*
↗3: *"Feynman Diagrams" on page 225*

↗4: *"Standard Model of Elementary Particles" on page 213*
↗5: *"Conservation Laws" on page 51*

along with the change of the involved particles. The weak interaction is the only one that is able to do so.

Heavy Bosons Make Interactions Weak

We know from other interactions that particles have to carry a specific charge to take part in the corresponding interaction. The weak interaction's charge is – surprise, surprise – the weak charge. And every particle is weakly charged, in other words: interacts via the weak force. So, fortunately, also the neutrinos interact via the weak charge. But as the weak charge is the only charge that neutrinos carry, the weak interaction is the only way for them to interact with other particles. And since the weak force is, in fact, not very strong, we do barely see them. Obviously, not with our own eyes, but also not via particle detectors where other particles would leave some electromagnetic traces. But what does it actually mean that an interaction is weak? Weak interactions happen "not that often". According to quantum mechanics, everything that can happen will happen within an interval of time. The corresponding probability depends on how strong the interaction is. And this strength is determined by a constant and the mass of the mediated boson. The weak interaction's constant is not that small, compared to other interactions. But the W- and Z-bosons are quite heavy – which is why the weak force deserves its name.

We need to take a look at the rules of quantum mechanics to get a qualitative answer of why a heavy boson causes a weak interaction. Imagine that the energy that would correspond to the mass of a W-boson according to $E=mc^2$ is not available during a process. This is actually true for most of the pro-

cesses of the weak interaction. In order to still make a process happen, you have to use Heisenberg's uncertainty principle ($\nearrow$6). Within a short interval of time you can "borrow" some energy for a W-boson and give it back when the W-boson decays. The amount of borrowed energy limits the time you can take it: the more you need, the shorter you can keep it. This means that the heavy W boson can only be borrowed for a very short moment. As the W boson cannot travel far in this short moment, the range of the weak interaction is also very short. This is in contrast to the infinite range of the electromagnetic interaction with its massless photon.

Elecroweak Unification

What is special about the weak interaction is not only the massive W-boson, but also the fact that also a second exchange boson exists: the Z-boson. It is massive as well, but not electrically charged. And as the W-boson, it couples to all the particles. So, if you see an electron and a positron attracting each other, this can be caused by either the exchange of

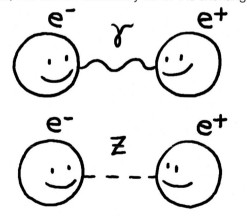

$\nearrow$6: *"Heisenberg Uncertainty" on page 151*

a photon (the exchange particle of the electromagnetic interaction) or a Z-boson.

You can see the two possibilities. Which one is happening? And how can you find out? Now, meet the weirdness of quantum mechanics once again. The answer to the first question is: both. And this also answers the second questions: you cannot. This example nicely shows the close connection between the electromagnetic and the weak interaction. In 1968, 35 years after the theory of the weak interaction emerged, the physicists Sheldon Glashow, Abdus Salam and Steven Weinberg managed to derive a unified description of the weak and the electromagnetic interaction. The unification refers to a common theory in which both interactions are just two different parts of a single interaction, the electroweak interaction. A next step would be to let the strong interaction join to an even bigger framework. The idea of such a common framework is named "Grand Unified Theory". So far, nobody has managed to find such a theory. But what has been solved recently is the question about the origin of the W- and Z-boson's masses. The answer is given by the Higgs mechanism (➚⁷), proven by the discovery of the corresponding Higgs boson.

Possible Transformations

One thing that we haven't discussed yet is: if the W-boson turns elementary particles into others, which ones can it modify? And into what? Concerning the leptons (➚⁴), the answer is simple: and an electron can be turned into an electron-neutrino and vice versa. The same is true for the muon and its neutrino, as well as for the tauon and its

neutrino. What is not possible is a change across these "lepton generations" – conservation laws (➚⁵) prohibit this. But for the quarks, things are different. Quarks are – in the same spirit as the leptons – also arranged in three families. We got to know the transitions within the first generation of up and down quarks as they occur in β-decays or processes of nuclear fusion (➚²). But, in contrast to leptons, quarks can also be converted across generations. They don't do it quite often, but they do. If we take a look at the up quark we see that it clearly prefers to be converted into a down quark. Maybe a strange quark. But a bottom quark? Meh. Physicists arranged these probabilities of all quarks into a matrix, the so-called CKM matrix (names after the physicists Cabibbo, Kobayashi and Maskawa).

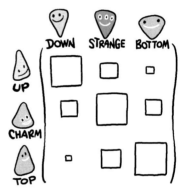

A precise measurement of the elements of this matrix – which are in fact not just boxes, but complex numbers – is of great interest. It allows getting an insight into a phenomenon called "CP violation" ("C" for charge and "P" for parity). It is responsible for the dominance of matter versus antimatter and hence for our existence. A good reason to be curious about it, isn't it?

➚⁷: *"The Higgs Mechanism" on page 241*

E=mc²

Energy and Mass – Almost the Same

E=mc^2: Energy equals mass times the speed of light squared. Mass and energy – without considering Albert Einstein's wise words, what would you guess as the relation between these two properties? Independent? Somehow the same? Or even contrary?

If I decide to go jogging after I ate a pound of candy, it feels as I have lost all my energy due to the increasing mass. Does the mass decrease my energy? Or was my energy converted into mass? Or can I, the other way around, convert half a kilo of belly fat into kinetic energy for a marathon? Well, such thoughts are tempting, but they have nothing to do with Einstein's famous E=mc^2. Whenever our body is moving or digesting food, it converts chemical compounds. This conversion, a re-arrangement of atoms, releases chemical energy only. All masses stay conserved (well, not exactly, but to a very good approximation). If you burn fat and your weight decreases, it is not the case that the fat is converted to energy and disappeared. After providing energy in chemical reactions, the atoms of your former fat molecules leave your body in different ways (for example via digestion and respiration). And there the mass goes!

If digestion has nothing to do with E=mc^2, then what does it deal with? It introduces a new type of energy. Usually we associate energy to a movement (kinetic energy) or talk about potential en-ergies. The latter concerns "potentials which can cause movements", such as potential energies like a hill with a certain height that a ball can use to roll down and gain energy. In E=mc^2 we have a new type of energy. Our ball would have this energy simply because it has a mass. There is nothing else that it has to do. If you want to clearly state that you are dealing with a rest mass and a rest energy, you can use an index 0: $E_0 = m_0 c^2$. As soon as our resting ball is moving, it also gains – next to the rest energy – kinetic energy. This kinetic energy is included in a factor denoted with γ:

$$E_{total} = \gamma \, m_0 \, c^2$$

The higher the velocity of a particle, the higher its kinetic energy, and the larger γ gets.

In our illustration you can see how γ – and hence the energy – increases with the velocity. The maximum velocity that can be reached is the speed of light.

Wow! You don't have to go to infinite velocities and make your energy explode. The speed of light, c, is enough. The theory of special relativity states that the speed of light, 299,792,458 meters per second, is constant and also the upper limit of everything that moves. The fact that the γ factor rises so steeply makes it very hard for particle physicists to accelerate particles to very high speeds. Scientists working at the world's largest particle accelerator, the LHC, know this problem best. Before the protons are injected into the LHC, they have an energy of 450 GeV. After the final acceleration, they reach 7000 GeV (or 7 TeV). While the protons' energies increased quite a lot, their velocity only shifted from 99.9997828% of the speed of light to 99.9999991% of the speed of light. Each bit closer to c costs a lot of effort. In some books you might read the term of "relativistic masses". People who use this expression include the gamma factor in the mass of a particle and use the simple $E=mc^2$ while saying that the mass increases. This might help to understand why you cannot reach the speed of light: The more massive you are, the harder it is to accelerate you. On the other hand, the expression of a relativistic mass is not well-defined. While the rest mass is always the same, the relativistic mass varies from the point of view. While a space ship traveling at very high speeds would get an increased relativistic mass from the point of view of a person at rest, it would not increase from the point of view of another spaceship traveling in the same direction at the same speed. For this reason, the expression of a relativistic mass is avoided. So

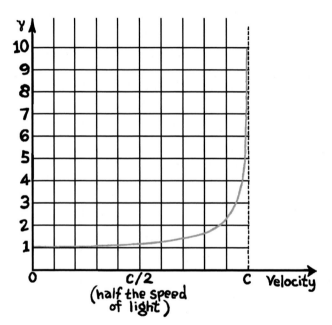

no worries: going jogging will not make you more massive.

What about the Massless Guys?

So far, so good. Nothing massive can get to the speed of light. And the faster you get, the higher your energy. But we have not thought yet about the guys that do already move at the speed of light. Always. It's light itself, represented by its quanta, the photons (↗¹). Photons are massless, but they do carry energy. Just think about a microwave oven (microwaves are the same type of electromagnetic waves as light, just with a higher wavelength) which heats up your food. Or sunlight that heats up your skin. This energy is proportional to the frequency of the photons. The higher their frequency, the higher their energy. The question is: do photons with their energy E behave like an object with a mass m, according to $E=mc^2$? This would imply that photons gain and lose energy when moving in a gravitational field. The famous Pound–Rebka experiment proved exactly that. Photons were sent up and down a 22.5 meters long tower of the Jefferson Lab at

↗¹: "Light" on page 7

Harvard's University and indeed: the gravitational field changed their frequency (see also ↗²). The fact that a massless energetic object behaves like a massive object also plays an important role in the explanation of the mass of … us human beings! While the Higgs mechanism (↗³) is made responsible for giving mass to elementary particles, our protons need much more than the Higgs mechanism to gain their masses. Read more about it in the chapter about the structure of the proton (↗⁴)!

to a larger nucleus. But the sum of the two small ones is larger than the mass of the fusion product. The difference in mass is converted into energy and released. This "mass defect" – and now your see what Maxwell thought of – is also used in nuclear power plants where nuclear fission (↗⁶) releases fractions of the masses of radioactive nuclei into energy. Next to fission, nuclei can also lose mass by simply emitting a massless photon, a so-called γ-ray (↗⁷).

While it is clear now that we can see mass being converted into energy, the question is: can we also do it the other way around? Can we create massive objects simply by converting energy into mass? Yes, indeed. You can read how we do that in the chapter about particle accelerators (↗⁸).

From m to E and Backwards

If E and m are somehow equivalent, can we go from one to the other? Yes, we can. One case where mass disappears is the decay of a particle called pion. It has a mass of about 15% the mass of a proton and decays into two massless photons which carry away the corresponding energy. Also the famous Higgs particle (↗³) can decay into two massless photons.

We can also use conversions from mass to energy to produce electricity. In nuclear fusion (↗⁵) two light atomic nuclei fuse

Let us end this chapter with an impressive example about E=mc². The detonation of the nuclear bomb "Fat Man" released an energy of about $9 \cdot 10^{13}$ J. An average sausage with a mass of 100 g has a food energy of about 300 calories. As one kcal corresponds to 4184 J, that sausage provides an energy of about 1.3 Megajoules. So in order to gain the amount of energy that was released by the explosion of a nuclear bomb, you would have to eat and digest about 72 million sausages. But if you could directly convert a sausages' mass of 0.1 kg into energy via E=mc², already 1% of the sausage would be sufficient. How does that sound?

↗²: *"The Theory of General Relativity"* on page 121
↗³: *"The Higgs Mechanism"* on page 241
↗⁴: *"The Structure of the Proton"* on page 245
↗⁵: *"Nuclear Fusion"* on page 175
↗⁶: *"Radioactive Decay"* on page 167
↗⁷: *"Alpha, Beta and Gamma Rays"* on page 171
↗⁸: *"Particle Accelerators"* on page 249

The Higgs Mechanism
Origin of Our Particles' Masses

Physicists try to quantify everything they can observe in nature. It helps them to derive the laws of nature. These laws often help to understand why many things behave the way they do. Sometimes physicists address their "why" to the very fundamental things, such as the properties of matter. Certain properties do not apply to the smallest building blocks of nature. An elementary particle has neither a color, nor a smell or temperature. These quantities simply cannot be defined for point-like, elementary particles. But there is one property that every particle has: a mass. You can measure it and use it to understand certain things: "Why is the mass of a helium nucleus four times larger than the mass of a hydrogen nucleus?" "Because the hydrogen nucleus consists of a single proton, while the helium nucleus is made of two protons and two neutrons. And as the mass of a neutron is about the same as the mass of a proton, it is in total four times more massive than hydrogen."

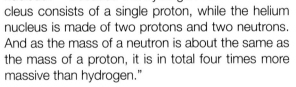

Elementary Mass

You can only play the game of saying "This object has a certain mass because it is composed of other objects with other masses" to a certain extent. Sooner or later you will end up with elementary particles. And even though they are point-like, which means that they have a diameter of 0 (the point can be delocalized as described by its wave function, ↗1), they have masses. While some are light, others are quite heavy. While the mass of an electron corre-

sponds to only 0.05% of the proton's mass, the top quark has the mass of 176 protons. Why is that? To be honest: nobody knows. We do not even have a clue. But while it is already very unsatisfactory not to know the reason for the individual masses, there is an even more fundamental problem with masses. When physicists write down their equations to describe our universe (↗2) they love to see a certain beauty in them. It might sound a little weird to relate a beauty to some equations, but maybe you have once experienced the feeling that some very complex procedure follows very basic rules. The beauty of the Standard Model lies in something that people call a symmetry – or to be more precise: a local gauge symmetry. The equations do not only follow these symmetries, but also allow to do very, very, very precise calculations and correct predictions of things that we can observe in nature. So everyone could be happy with them, right?

As Always: Mass, the Big Problem

If you think that everything is fine then often one guy comes along that ruins everything: mass. Tasty food ends in mass. Mass ends in problems. In very extreme cases, the addition of mass can even lead to some symmetry, namely a rotational symmetry (in case one turns into a sphere).

But what is the situation if you start with the equations from the Standard Model, leave all particles massless and then suddenly add a mass to them?

↗1: *"Wave-Particle Duality" on page 143*
↗2: *"Standard Model of Elementary Particles" on page 213*

Instead of introducing an additional symmetry, you destroy the beauty of the equations' symmetry, the local gauge symmetry. To explain this in detail would go far beyond the scope of this book. But so far we can conclude: Beautiful, well working equations which can calculate everything we observe (except gravity) do not allow particles to have masses. The question is: What to do?

Higgs: Solving the Problem

Well, you cannot deny the particle's masses, so you have to come up with an idea. And some people did: in the 1960s the physicists Englert, Brout, Higgs, Guralnik, Hagen and Kibble developed the idea of an extra field which had to be added to the Standard Model's equations. Today, this field is often simply called Higgs field. Commonly known fields are for example electric and magnetic fields (↗3). Such fields are called "vector fields". They assign a vector to each point in space. This vector has a magnitude and a direction it points to. So for our example the vector of a magnetic field tells us the strength and the direction of a magnetic field. You can also imag-

ine a "river field" which describes the flow of water within a river: how fast does the water move and in which direction? As this can be different at different places, you need a vector for each point in space (or more precise: in the river). In contrast to vector fields, our new Standard Model field was proposed to be what is called a scalar field. Here, each point in space simply gets a number, no direction. Take a temperature map as example: How warm is each point on a map? So "Temperature" can be described with a scalar field.

Now, what is the use of this Higgs field? It is scalar, so it has no preference in direction. It is simply everywhere. And it does two things with our Standard Model equations. Firstly, it couples to those bosons that mediate the electromagnetic and weak interactions (↗4). This coupling causes something that physicists call a "spontaneous symmetry breaking". As a result, the equation's beauty and initial symmetry (without masses) is preserved but they still get some terms that look like masses of the bosons. So you introduce the boson masses indirectly by couplings to the Higgs field.

↗3: *"Light" on page 7*
↗4: *"The Weak Interaction" on page 233*

While this symmetry breaking is only hard to imagine and concerns only the masses of the bosons, we can now take a look at how the Higgs field gives masses to fermions, our matter particles. What we call a mass is – according to Newton's laws – a resistance against acceleration. The smaller your mass, the easier it is to accelerate you. Now imagine that all the massless matter particles in the Standard Model interact with the new scalar field. Some do more, some do less. We call this amount of interaction a "coupling strength". Take a look at the illustration, showing a bunch of students. The students represent the Higgs field. Two guys try to reach the coffee machine on the right. One is a friend of Maxwell. Nobody knows him, and that's an advantage, at least concerning the time it takes him to get coffee. We can say that his coupling strength to the Higgs field is low. As he can move almost freely, we can say that due to his small coupling strength to the Higgs field he moves as if he had a low mass. That's not the case for poor Maxwell: He, as the superstar of this book, is so famous that everybody wants to talk to him, interact with him. That slows him down. So this time: Large coupling strength, looking like a large mass.

And that is exactly the way that fermions get masses: they interact with the Higgs field. And as the in-

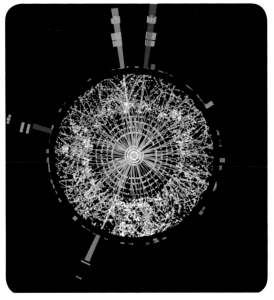

teraction strength is different for each particle, the masses are different as well. Developed in the 1960s, this has simply been an idea for a long time. The explanation of masses could also work completely different. But in 2012, the ATLAS and CMS experiments based on the world's largest particle accelerator LHC (↗[5]) made an interesting observation. By using the energy that the protons accelerated in the LHC gained, a particular particle could be produced. It decays immediately (↗[6]) and can only be observed indirectly via its decay products which leave traces in the particle detectors (↗[7]) ATLAS and CMS. You can see one of the snapshots of that particle's decay products on this page. Many of such reconstructed particles were needed to really confirm its existence. That is because some other particles, which were already known, can decay in a similar way. But it was indeed the discovery of a new particle: the Higgs boson! This new boson was predicted in the theory of the Higgs field. It is an excitation of the field itself, as if the students would cluster without Maxwell being there. The Higgs boson's discovery could finally, 50 years after the idea of the Higgs mechanism was made, prove that theory. And one year later, in 2013, the explanation of the mass generation mechanism was awarded with a Nobel prize for the physicists Englert and Higgs. One less mystery in the universe.

↗[5]: *"Particle Accelerators" on page 249*
↗[6]: *"Particle Decays" on page 221*
↗[7]: *"Particle Detectors" on page 253*

Image: ATLAS Experiment © 2014 CERN

The Structure of the Proton
Or How to Get Mass without a Higgs

It is always tempting to blame someone for one's own problems. If the bathroom scale shows too much, then whose fault could that be? The chocolate's? The fast food's? Maybe the own laziness' fault? Oh wait! There is this Higgs boson! And its Higgs mechanism is responsible for giving all particles their masses. So: No Higgs, no being overweight, right? Well, not really …

It is not the case that the Higgs mechanism takes a look at Erwin, Maxwell and Emmy and says: "I would give you this and that mass". Only elementary particles get masses. In the chapter "Atoms vs. Elementary Particles" (↗[1]) we have seen that only the smallest particles which cannot be broken into anything smaller can really be called elementary. So let us play the game of "breaking and checking what is inside" once more, but this time with a focus on the particles' masses. When reaching microscopic scales, it is convenient to use a different unit for masses than kilogram or pound. Particle physicists use MeV ("mega electron Volts") where one MeV corresponds to $1.7 \cdot 10^{-30}$ kg. You see: it is a really tiny unit, made for tiny particles. Actually, MeV is a unit of energy, but it can also be used for masses due to the equivalence of mass and energy (↗[2]).

The Proton – Elementary or Not?

Let us start with a human being with a mass of 70 kg. He consists of $1.5 \cdot 10^{27}$ oxygen atoms with a mass of 14904 MeV, $9.8 \cdot 10^{26}$ carbon atoms with

a mass of 11178 MeV, $3.9 \cdot 10^{27}$ hydrogen atoms with a mass of 938 MeV, $6.0 \cdot 10^{25}$ nitrogen atoms with a mass of 13041 MeV, $1.6 \cdot 10^{25}$ calcium atoms with a mass of 37260 MeV, $1.2 \cdot 10^{25}$ chlorine atoms with a mass of 33021 MeV, $1.4 \cdot 10^{25}$ phosphorus atoms with a mass of 28852 MeV and some minor contributions from other atoms. If we sum it all up, we reach the 70 kg. Great!

Now we go one step further, looking at the structure of one of these $6.5 \cdot 10^{27}$ atoms in our body. Take carbon, for example. Its mass is about 12 times the mass of hydrogen. We know the structure of atoms, represented by Rutherford's atomic model. Almost the whole mass of an atom is focused in the nucleus of an atom. These nuclei are made of protons and neutrons, which have almost the same weight. And as a carbon nucleus is made of six protons and six neutrons while a hydrogen nucleus is just a single proton, this perfectly explains the 12 times higher mass of carbon. Great!

Well, almost. This time we used a little trick and swept something under the rug. Twelve times the mass of a proton is not exactly the mass of carbon. As the nucleons, namely the protons and neutrons, have to stick together, they need some binding energy. This binding energy, which is subtracted from the energy corresponding to the sum of the nucleon masses, leads to the little mass difference (see the explanation about the mass defect, ↗[2]). So the carbon atom is slightly lighter than twelve hydrogen

↗[1]: *"Atoms vs. Elementary Particles" on page 205*
↗[2]: *"E=mc²" on page 237*

atoms. But the effect is marginal, just about 0.8%. You see that in the nuclear regime, mass is no longer only massive, but slightly modified by energies.

Let us now go deep into the jungle of elementary particles. Going one level deeper brings us from the proton to its constituents: quarks. And these guys are – as far as we know – fundamental, getting their mass from the Higgs mechanism. Physicists got to know that there are quarks in the proton by performing scattering experiments. We learned about them in the chapter "Atoms vs. Elementary Particles" (↗¹). In the Rutherford scattering experiments, α particles (↗³) were scattered off atomic nuclei to reveal the secrets of the atomic structure. It is like shooting marbles on a target to guess its structure from the scattering directions. And in that chapter we have also seen a picture of a giant electron gun which was targeting protons. Such scattering experiments unraveled the proton's structure. At low energies, the protons behaved like a ball. But for higher energies one could see that indeed the electrons were

scattering off little quarks inside the proton: Two up-quarks and one down-quark.

By increasing the energy of the electron gun even further, one could even kick out one of the quarks from the others. But then, something strange happened. It is the strong interaction (↗⁴) which makes the quarks stick together to a proton. And the energy in the field of the strong interaction gets larger for larger distances. So large that if you try to separate a quark from its friends, the energy between them will increase and increase until via E=mc² another pair of a quark and anti-quark is created. So in contrast to all other particles that built up our matter, quarks cannot be split off. You can separate an electron from an atom, a proton from a nucleus, but a quark from others? No.

So in a simple picture, a proton consists of three quarks. If you look closer, you can see that gluons (represented by the little springs) glue the quarks together. And if you look even closer, the gluons do funny things while traveling from one quark to another. They split into a pair of a quark and anti-quark. Just for a short amount of time. We got to know such temporarily created particles as "virtual particles" in the chapter about Feynman diagrams. If you would zoom

PROTON

electron

HIGHER ENERGY

QUARKS

↗³: *"Alpha, Beta and Gamma Rays" on page 171*
↗⁴: *"The Strong Interaction" on page 229*

in even further, you could see the virtual quarks again exchanging gluons, and so on, and so on. It is an infinite repetition. So that is all about the structure of the proton. But what about its mass?

It's All about E=mc²

Let us zoom out to the basic proton picture: Three quarks and gluons. The gluons are massless, so let us take a look at the quarks. Both the up- and down-quark are lighter than 10 MeV. This is actually the mass they gain via the Higgs mechanism. But three times 10 MeV is … not at all 938 MeV, which is the mass of the proton.

This magic can again be explained by E=mc². But this time, it causes more than just a 0.8% correction. It makes up almost the whole proton mass! The Higgs mechanism is only responsible for a ridiculous amount of a few percent, namely the quark masses. But which energy is here causing the proton mass?

We have to take a look at the Heisenberg uncertainty principle (⬈5). It states that the more precise you know a particles' position, the less you know about the value of its momentum. At macroscopic scales, this effect does not play a role. Of course, the definition of "precise" is relative. Some people might say: "I know precisely where my car is. In the garage. And I also know precisely what its momentum is: zero. It's turned off." True. Within an uncertainty. Surely, you will not be able to determine the position of your car up to 1 femtometer, so 10^{-15}

meters. But this is exactly what we claim to know about the position of the quarks which are trapped inside a proton. At these scales, the uncertainty principle really plays a large role. The uncertainty of the momentum grows significantly. This means that the quark cannot sit around quietly, it will shake violently, extremely fast and outraged.

So the quarks will move around inside the proton very fast. But still, they are not allowed to escape it due to the special property of the strong interaction. Via their fast movement, the quarks carry a lot of kinetic energy. And the quarks share this energy with the gluons they exchange. So the proton is basically just a small packet of energy which cannot be released, and this energy is what we observe as the proton's mass. E=mc² in a box.

⬈5: *"Heisenberg Uncertainty" on page 151*

"We've seen some surprising results from particle collisions, but never one quite like this."

Particle Accelerators
Time Machines and Big Bang Creators

They are said to be able to create black holes. They work as time machines, bringing us to the beginning of the universe. Some people claim that they are able to re-create the Big Bang. Giant machines, built in tunnels underground: Particle accelerators. Why are Physicists building such machines and how do they work?

The name says it: particle accelerators speed up small particles. Once they are fast enough, they are smashed against each other (particle collider) or against something standing still (fixed target accelerator). There are two main reasons why one should do things like that. The first one is quite obvious. If you smash something and it breaks, you can see what is inside. If you like to test this principle, you can build your own piggy bank accelerator to see what's inside.

Physicists are more interested in tiny particles, for example protons. They do it because they don't know a lot about a proton's ingredients. Particle accelerators used for this purpose can provide a lot of energy that is needed in order to break a proton. You cannot simply use a hammer and break it.

Converting Energy to Matter

There is another reason to build particle accelerators next to looking inside objects. Once you have accelerated particles to a very high speed, they will have a lot of energy as well. And in the process of collision, this energy is released. Think of a cannon ball hitting a wall. The faster it is, the more energy it has to...let's say "modify" the wall. If you collide two particles after they were accelerated to very high speed and thus energy, there is nothing the particles can do with the energy if they simply hit each other. If they are not elementary particles ($\nearrow^1$) they will break. In the case that they are elementary, they cannot break. But they will also do something else. Something that we cannot observe in our macroscopic world. They will make use of Albert Einstein's famous $E=mc^2$ and convert their energy into new matter! And even antimatter ($\nearrow^2$). Both will always be produced in equal amounts, mostly in pairs of matter and antimatter. You can produce something that is well known and that is available in nature. But you can also create something very fancy like a strange quark (that's their real name) and an anti-strange quark. The heavier these particles are, the more energy you'll need to create them. The Standard Model ($\nearrow^3$) includes all known building blocks of matter (but we know that even more should exist). Right after the Big Bang ($\nearrow^4$) they all existed. They were continuously created and decayed ($\nearrow^5$). But as the universe evolved and cooled down there was a time when suddenly not enough energy was available to re-create some of these particles. That's why today we don't see them in nature. But we can build particle accelerators with

$\nearrow^1$: *"Atoms vs. Elementary Particles" on page 205*
$\nearrow^2$: *"Antimatter" on page 217*
$\nearrow^3$: *"Standard Model of Elementary Particles" on page 213*
$\nearrow^4$: *"The Big Bang" on page 97*
$\nearrow^5$: *"Particle Decays" on page 221*

enough energy to see how things were right after the birth of the universe: which particles did exist and how did they interact?

Energies of Particle Beams

You may ask why people accelerate elementary particles and not something like tennis balls. There are quite a few talented people that really know how to accelerate tennis balls. So why deal with small particles? Think of smacking the neck of a friend with your bare hand. It hurts. Now think (and please, only think!) of smacking his neck again with the same energy, but this time with a little needle placed on your hand. It hurts. A lot more. Why? You had the same energy, but this time you have focused it to a very small point. The "energy density" at the needle is much higher. As the smallest things we know are little particles, these are the ones being accelerated if we want to have the highest energy densities.

Let us get a feeling of the energy of a particle which made its way through a modern particle accelerator. The biggest one today is the "Large Hadron Collider", LHC. It is located at the research facility CERN close to Geneva, Switzerland. It accelerates protons up the 99.9999991% of the speed of light. This is really fast, especially if you consider that the laws of Einstein's theory of special relativity ([6]) do not allow anything to be faster than the speed of light. Can you think of anything you have ever seen that has the same amount of energy? No? What about a mosquito? If it flies about 2 cm/s (not that fast) it can already compete. This sounds completely crazy. But you should consider that the energy of the mosquito is spread to about 10^{21} protons. Again, energy density makes the difference.

How to Accelerate

Now it is time to learn how such accelerators work. My Grandma knows a lot of things. And by chance she also has a particle accelerator at home. It is nothing fancy but simply an old TV screen. One from the time before they were flat. It works the following way: In the back of the TV there is a little wire which is heated by an electric current. Due to the heat, electrons moving through the wire get sufficient energy to be released from the wire and to move freely. What you need to do is to make them move into the right direction (towards the screen, to make the colored dots glow and produce a nice picture). You cannot simply drag an electron. But the fact that it is a charged particle allows to accelerate it with an electric field. It will move towards a more positive electric potential. The potential differences are called voltages and the higher the voltage, the higher the speed of the electron will be in the end. That's why physicists quote the energy of accelerated particles in "electron volts" (eV). A particle which was accelerated by a voltage of 1 Volt has the energy of 1 eV. What works for an electron works also for a proton (it just flies in the opposite direction) or any other charged particle. So if you want to accelerate a

I CAN REALLY PICK UP SPEED LIKE THIS

...BUT ITS GOING TO TAKE A LONG LINE OF TUBES!

TO COLLIDER

[6]: *"Relative Space and Time" on page 117*

neutral atom you first have to rip off one of its electrons to make it being charged. Super high energies require super high voltages. There is a limit which you cannot get above for technical reasons (about 30 million Volts). Is that the end of all acceleration limits? No! Just be smart. Instead of one big kick you can give several small kicks to a particle:

We see a proton and a negatively charged tube. The proton gets accelerated. The following tube has the opposite charge. In case the proton comes out of the first tube it will get repelled by the second. Too bad, no acceleration. For that reason the voltage needs to switch while the proton is in the first tube. It continues on in this way. So instead of applying higher voltages you simply need more tubes. The problem with this is that your accelerator gets larger and larger. Such "linear accelerators" can be even a few kilometers long, as for example the Stanford Linear Accelerator in Menlo Park, California.

Going in Circles

If you don't have enough space or money to build a more powerful accelerator you can think of alternatives. What about having the accelerated particle come back to the entry of the tube that first accelerated it? So instead of many tubes you just have to make it so that it goes many rounds. After kicking a particle with an electric field it will go straight and not come back on its own. To bend it you can use magnetic fields. You will need a lot of powerful magnets to build such a circular collider. The largest particle accelerator in the world, the LHC, needs 1232 magnets to make protons turn around at a circumference of 27 kilometers. And the mag-

nets are powerful. Their magnetic field is created by coils with a current of 12,000 Amperes going through them. Your switch at home blows if more than 16 Amperes go through your cables. Higher currents would make your cables get hot and melt. The cables in the LHC do not melt because they are cooled down to -271 °C (really cold!). At this temperature, the special material they are made of gets "superconducting" ($\nearrow$7) and the cables won't heat up any more.

The LHC is built in a tunnel about 100 meters under ground. This is not because the machine is dangerous or needs to be hidden, but it's simply cheaper to dig a tunnel than to buy all that land on the surface. Oh, what we should not forget: Next to magnets, electric fields, and a tunnel, we also need something else. Imagine you are driving on the highway at 99.9999991% the speed of light. In the case that something slow appears in front of you, you will have a serious problem. The situation is similar for particles in an accelerator. So what you should not forget is a good vacuum ($\nearrow$8) in your beam pipe.

Two beams of protons are sent through the accelerator rings of the LHC. Both beams consist of 2802 bunches with 10^{11} protons each. There are four interaction points within the 27 km long accelerator in which the two beams are crossed and the proton bunches collide. Every second 40 million proton bunch collisions occur, releasing their energy into pairs of matter and antimatter. Giant particle detectors ($\nearrow$9) make pictures of these collision events and allow to analyze the data. And we don't know which new particles the LHC might find. Maybe supersymmetric particles ($\nearrow$10)?

$\nearrow$7: *"Superconductors" on page 179*
$\nearrow$8: *"Vacuum and Air Pressure" on page 23*
$\nearrow$9: *"Particle Detectors" on page 253*

$\nearrow$10: *"Supersymmetry" on page 299*

Particle Detectors
Showing the Invisible

Hearing the word "detector", what do you think? It sounds a little like a detective. Someone who wants to find out what actually happened. Maybe you already encountered a detector, for example at the airport. Metal detectors help the personnel to find metal materials you might carry with you and which cannot be seen by eye. While you can see metal objects once the detector points them to you, some things will stay forever unwitnessed by your naked eye: elementary particles. They are simply far too small to be seen directly. So the question is: how can you make invisible things visible in a detector?

The Charged Ones Cannot Escape Unseen

What do airplanes and wild animals have in common? They often pass by without being seen. But we know that they were there when they left some traces. A typical animal trace appears in the snow. If you know the type of trace you can directly infer back to the type of animal: cat or horse, you will know it.

A typical trace of an airplane looks like a long and thin cloudy strip that appears in the sky. Let us consider the sky as an airplane detector for a moment. We can use this example to explain a very common particle detector which is neither very powerful, nor very modern or fast. But it can make tiny little par-

ticles like α-rays, β-rays (↗[1]), protons and muons visible. Its name is "cloud chamber". It consists of a closed glass box in which alcohol is heated up. The alcohol evaporates and fills the air inside the box.

One says that the air is saturated with alcohol vapor. While the heating of the alcohol takes place at the top of the box, a very cold metal place is located at the bottom. The saturated air at the bottom gets cooled and as a consequence of the cooling, it can keep less alcohol. So this time the air is not only saturated, but "oversaturated": it would really like to get rid of the alcohol. If you can see your own breath in the cold winter air, that's precisely the same effect.

To allow the alcohol to be squeezed out, the air needs a trigger. Something that creates a place where the alcohol molecules can attach themselves to form droplets. Such a trigger is a charged particle coming along, which carries enough energy to ionize the air in the cloud chamber. Then the alcohol vapor will condensate on the ionized atoms exactly along the path that the charged particle took. And there you go: you made the invisible charged particle visible!

If you carefully check the shapes of the traces in a cloud chamber, you will see that some are thick and short while others are thin and long. Of these thin lines some are quite straight while others make zigzag moves. The different trace types come from different particle

---

↗[1]: *"Alpha, Beta and Gamma Rays" on page 171*

types that cause them: The more charge they carry, the more they ionize, the more vapor condensates and the broader the traces get. That's why α-particles with their double positive charge leave broad traces. And as these double-charged α-particles ionize a lot, they lose a lot of energy. And the more they lose, the faster they are stopped. This makes their traces so short. You can also see differences between electrons (β-particles) and muons: the heavier muons get less distracted by scattering off the air molecules and have hence straighter tracks than electrons.

The principle of a cloud chamber is the same as for the airplane traces. There, hot air coming out of the turbines gets cooled in the sky and the water vapor condensates at little dust particles. We can now ask ourselves: What should a particle detector tell us about a particle that it detects? While there are also very sophisticated detectors which provide very sophisticated information, the main questions we would like to address when hunting the fundamental building blocks of nature are: where did the particle go? And which energy did it have? The detector types addressing these questions are tracking detectors and calorimeters.

Tracking Detectors – Tell Me Where You Went

When you follow an animal's path by checking its traces in the snow, you do not want to disturb the animal, do you? Tracking detectors want to do the same with the particle they track. The tracking itself works similarly as in the cloud chamber. A charged particle enters a detector material and ionizes it.

This means it kicks out electrons off atoms and splits neutral atoms into negative electrons and a positive remnant. If you apply a voltage at your detector, you can make the positive charges going to the negatively charged side (the cathode) and vice versa. This allows you to measure a voltage pulse caused by the ionized atoms. You can digitize this electronic signal and get something like a "click". The illustration shows you a very modern implementation of this method via a semiconductor diode. If you add a voltage to such diodes, no current flows until you add some extra energy, in this case by a traversing particle. The good thing about such a semiconductor detector is that it can be built really small, as we do in microprocessors, for instance. Such very small structures are exactly what you want. Every time a particle traverses your detector you can say "it was somewhere in here". And the smaller your detecting material is, the more precisely you know where your particle went. Modern detectors allow to determine the position of a particle with a precision of a few micrometers. Instead of a semiconductor diode you can also use a tube filled with gas. The famous Geiger counters (the ones that go click-click-click when detecting radioactivity) work like this. Cerenkov detectors, which were introduced in chapter "The Neutrino" (↗²), are another way to make the path of particles visible by emitting a light cone instead of releasing charges via ionization.

electron

↗²: "The Neutrino" on page 209

The information about the path of a particle is of particular interest if you put a strong magnetic field around it. Charged particles get bent on a circular orbit. The faster they are (more precise: the more momentum they have), the larger their bending radius is. And the direction of the bending is determined by the sign of the particle's charge. So if you precisely know the path of a charged particle in a magnetic field, measured by a tracking detector, you can tell about its charge and momentum.

Calorimeters – Stopping Particles

While tracking chambers try not to deflect particles too much, there are other detectors working exactly the opposite. They try to stop particles violently. It is like as if they hit a wall and lose all their energy. Such detectors are called calorimeters. The important thing is that they do not only want to stop the particles, but also to measure all the energy they lost. This way you know how much energy they had. Calorimeters can also be used to determine the energy of uncharged particles, such as neutrons or photons. You could not achieve such a measurement with tracking detectors since neutral particles do neither ionize, nor get deflected.

So how does a calorimeter work? Many of them are arranged as a "sandwich calorimeter". A passive layer enforces a reaction between the particle and the detector material. The active material, following the passive, reads out the energy that the particle lost in such a process. Let us see how a calorimeter works if it measures the energy of an incoming photon.

It converts into a pair of a positron and an electron (↗3). The electrons and positrons emit high energetic X-ray photons. They again split into electron/ positron pairs and the story goes on. The average energy per particle decreases and in the end you count the collected energy of the very low energetic photons and electrons, which are read out in the active layers. Such "avalanches" of electrons and photons are called "particle showers". It is easy to imagine that an electron would start the same type of shower, simply starting with the emission of a X-ray photon. Hadrons can also be trapped in a calorimeter by losing their energy in showers.

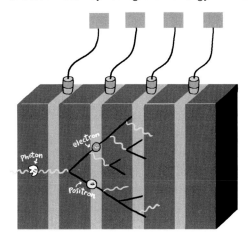

Such showers have a different structure and are caused by processes of the strong interaction (↗4).

All Together

You can arrange tracking detectors, magnetic fields and calorimeters to powerful general purpose detectors, such as the ATLAS and CMS detectors (↗5). The ATLAS detector for example has 100 million readout channels (such as a 10 MPix camera), 3000 km of cables and can analyze up to 40 million pictures of particles traversing it.

↗3: *"Antimatter" on page 217*
↗4: *"The Strong Interaction" on page 229*
↗5: http://atlas.ch and http://cms.web.cern.ch

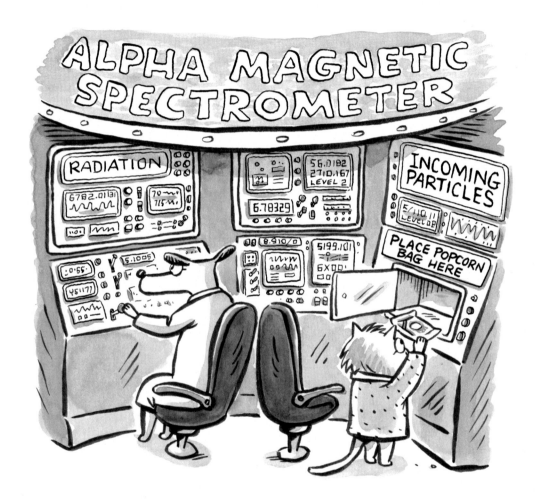

Cosmic Radiation
Sent from Unknown Accelerators Far Away from Us

Radiation is everywhere. We do not feel it, but sometimes we are at least aware of it. If a cell phone call reaches us, for example. The information of the call is transported via electromagnetic radiation. There is still a large debate whether this electromagnetic radiation harms us or not. But there is one special type of radiation which does for sure at least have the potential to harm us: ionizing radiation. This radiation has enough energy to kick electrons out of atoms and to break chemical bonds. We can quantify the radiation that is exposed to use by its dose which quantifies the amount of energy that the radiation deposits in our body (which can potentially damage our cells). The total dose that reaches us every day is half natural and half man-made. Man-made ionizing radiation is what we could avoid by not creating it. But one should know that about 90% of the man-made radiation comes from medical diagnosis as X-ray and computer tomography images. In case of an emergency, that's probably something you would not deny.

The natural radiation is dominated by radon gas (see nuclear decays for details) which originates from uranium decaying within the Earth's crust. In the beginning of the 20th century, scientists wanted to investigate this natural radiation and measured it in various places. They checked in particular also the dependence on the height above the ground. If all the natural radiation stemmed from the Earth's crust, it would decrease the further away you are from the ground. The hypothesis was tested and … failed. Very detailed studies which opened the door to the understanding of a new source of natural radiation were performed by Victor Hess in 1911 and 1912. He put some ionization measurement devices onto a balloon and measured a decrease of radiation up to about a mile. But then, instead of decreasing further, the amount of radiation increased. At 5000 m it was already twice as intense as on the ground, increasing to up to 40 times the ground intensity at 9300 m.

Radiation from Above

Hess concluded that there was a new type of radiation coming from far above, hitting the atmosphere of the Earth and then making its way down to the ground. The atmosphere shields the Earth against this radiation, and the higher Hess' balloon flew, the thinner the atmosphere, and therefore the less shielding was there. So the more radiation he could measure. For the discovery of this new so-called "cosmic radiation" Hess was awarded the Nobel Prize in physics in 1936.

Cosmic rays are presently still a topic of large interest. Two closely connected questions puzzle physicists: 1. What is the source of the cosmic rays? and 2. what accelerated them to such enormously large velocities? So far, we have not talked about the velocity and the related energy. The energy of cosmic particles hitting the Earth's atmosphere reaches up to 10^{20} eV. This is more than 14 million times the energy of protons accelerated in the Large Hadron Collider (LHC), the largest particle accelerator ever built by mankind. It seems as nature has much more powerful ways to accelerate particles than we do. And as we are curious, we would like to find out how nature does it.

From a Particle to a Shower

These super highly energetic particles do not make it down to the Earth's surface and we cannot investigate them directly. Instead, we use indirect methods, which use the processes which are illustrated in the figure on this page. A cosmic particle – in most of the cases a proton – hits the first layer of the atmosphere. It then collides with a particle from the atmosphere via the strong interaction (↗1). Part of the primary proton's energy is converted into new matter and antimatter (↗2). The newly created particles induce further reactions with other atoms of the atmosphere. After each reaction, the number of produced particles increases. Simultaneously, the average energy per particle decreases. Such particle cascades also occur in particle detectors (↗3) called calorimeters. They are called hadronic (if induced by a process of strong interaction) and electromagnetic (if induced by the electromagnetic interaction) showers. The energy of the initial cosmic particle determines the spread of the shower. One particular type of particle induced in such showers barely interacts with the atoms of the atmosphere: the muon. As it loses almost no energy while traveling towards the ground, it can easily reach the surface. We can see tracks of these cosmic muons in cloud chambers (↗3), count them in self made muon detectors made out of coffeepots (↗4) and see them passing through particle detectors even a hundred yards underground (such as ATLAS and CMS, ↗5).

it's raining rays

Studying cosmic rays requires measuring their energy and their direction. To do this, you have two options. Either you directly measure the cosmic rays before they hit the Earth's atmosphere and convert into showers. For that, you have to measure it like Erwin and Maxwell and go directly into outer space. The

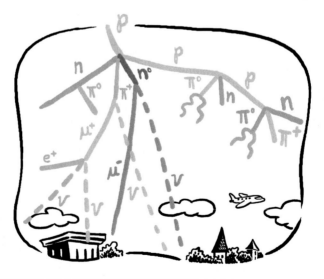

↗1: *"The Strong Interaction" on page 229*
↗2: *"Antimatter" on page 217*
↗3: *"Particle Detectors" on page 253*
↗4: http://kamiokanne.uni-goettingen.de
↗5: http://atlas.ch and http://cms.web.cern.ch

AMS experiment (↗6), located at the International Space Station ISS, does this. Another way is to measure the showers on the Earth's surface. If you want to study the energy of the initial rays you have to measure the size of the showers: the larger the energy, the larger the shower. Really high-energetic rays can induce showers with several hundred acres of size. Nobody could build a detector which is large enough to cover it. Instead, scientists build several small detectors, let each one detect a small piece of the shower and then have them communicate in order to put the pieces together as one big picture. Such detector arrays consist either of telescopes which detect the photons of the showers – like the HESS, MAGIC and VERITAS Experiments – or of Cherenkov detectors (see the neutrino chapter for details) which measure charged particle components of the showers. An example for such a Cherenkov detector array is the Pierre Auger observatory (↗7), designed for the very large cosmic ray showers. It consists of 1600 detectors, spread over 3000 km² in the Argentinian countryside. The detectors consist basically of a water tank, a solar panel and an antenna. You can see it in the picture.

Antenna-Loving Birds

There is a funny story about it: Physicists started to wonder why more and more detectors ran out of energy. It seemed as if the solar panels stopped charging them. Take a look at the detector in the photo. Do you have an idea? The antenna, located right above the solar panel, is a very comfortable place for birds. And from time to time, birds have to drop something. After a while, the solar panels got covered with bird droppings. Fortunately, there was an easy solution to the bird problem: cleaning the solar panel and turning the antenna by 180 degrees.

Cosmic ray studies are still ongoing. We still would like to know how nature manages to accelerate particles to such high energies. One assumption is that supernovae explosions (↗8) accelerate particles which surf on their shock wave fronts. Another reason to study cosmic rays is to understand the radiation that astronauts are exposed to, in particular in the context of a future travel to Mars. The used aircraft do not have a protective shield such as the Earth's atmosphere. And by the way: you do not have to travel to Mars to see the effect of an increasing radiation that you are exposed. Just take an intercontinental flight. The higher your plane, the less protective shielding you have. The effect is small if you do not do this every day. But personnel working on a plane has to keep an eye on their total accumulated dose of radiation.

↗6: http://www.ams02.org
↗7: https://www.auger.org
↗8: *"Supernovae"* on page 83

Image: Pierre Auger Observatory

Neutrino Oscillations
Particles Changing Personalities

Neutrinos are really funny guys. They do barely interact with anything, don't carry a charge and have no mass, even though they belong to the matter particles. No mass? Okay, almost no mass. In the mean time physicists know that they must have a least a very small mass to be able to do something which is known as "neutrino oscillation". This effect is hard to believe and we will soon get to know how it works. But let us start at the origin of the neutrinos' strangeness.

The biggest provider of neutrinos reaching us here on planet Earth is the sun. We say that about 70,000,000 neutrinos per second pass every cm^2 of our body surface. As the neutrinos do not interact with us, how do we know that there are so many? What scientists did was calculate a model of the sun. How large is it, how much energy does it release? Which processes take place inside? All this information is included in the so-called "Standard Solar Model". Many of the processes within the sun involve nuclear fusion ($\nearrow$1): two protons combine and form a deuteron, after that one of the protons transforms into a neutron via a weak interaction ($\nearrow$2) process. The positive charge of the proton is "taken away" by a positron, which is emitted together with an electron-neutrino. Most of the solar neutrinos reaching us stem from these fusing protons. So far, so good. People cal-

culated the number of neutrinos which should reach us and compared it with a measurement.

Counting Neutrinos

Of course, this first measurement of the solar neutrino flux was not that easy. A first successful attempt was made in the Homestake mine in South Dakota. Protected from other radiation sources at almost one mile below the surface, about 100,000 gallons of Tetrachloroethylene were used for neutrino detection. You have never heard of Tetrachloroethylene? No worries. All we need to know is that it contains a lot of chlorine. And that is what the solar electron-neutrinos like: they convert the chlorine into argon, a light gas. The argon atoms produced in this reaction are on the one hand very light, will try to get out of the detector and can be captured on their way. On the other hand, the special argon isotope produced in this reaction will decay radioactively. This makes it easy to detect and count the argon atoms. Don't expect too many of them. Every second day, the physicists running the experiment counted one of them. Not many, given that there were more than 2,000,000,000,000,000,000,000,000,000 chlorine atoms in the experiment. You see: counting neutrinos is not an easy thing.

$\nearrow$1: *"Nuclear Fusion" on page 175*
$\nearrow$2: *"The Weak Interaction" on page 233*

So much effort of exhausting atom counting was made just to confirm the solar neutrino model. And it did not succeed. Only about one third of the number of the expected neutrinos coming from the sun were observed. This immediately raised the question: whom to blame? The experimentalists (such as Maxwell) counting the neutrinos? Or the theorists (like Erwin) who calculated the expected number of neutrinos?

What Happened to the Neutrinos?

Other theorists and also experimentalists tried to do better. But they all came to the same conclusion: The neutrinos must have disappeared on their way from the sun to Earth. So  the theorists started to think of new ideas about how the electron-neutrinos from the sun could have disappeared. One idea that seemed rather obscure in the beginning was that the neutrinos could oscillate. We know that there are three types of neutrinos: electron-neutrinos, muon-neutrinos and tau-neutrinos. What if they could convert into each other? This is what the oscillation theory states. Let us think about the consequences of such an oscillation in our daily life. Imagine Maxwell fills his whole fridge with 20 bottles of his favorite apple juice. The next day he finds one bottle of orange juice in his fridge which he never put in there. A few days later, half of his apple juice turned into orange juice. And after waiting a little more, there is only orange juice left. Okay, we all know similar "magic" going on in our fridges when nice food turns into mold. While there is nothing special about that, Maxwell's oscillating orange juice will turn back into apple juice! The oscillating character of this process is that it repeats again and

again. The oscillation of particles can be explained mathematically in the context of quantum mechanics where particles are described by propagating waves. The wave functions of different neutrino states oscillate in a similar way as two coupled pendulums oscillate which can be nicely illustrated (↗3).

A neutrino oscillation could be proven if one would not simply observe missing electron-neutrinos, but also observe additional muon-neutrinos coming from the same direction. To clarify this, more advanced experiments had to be performed. The Homestake experiment was not able to say anything about the direction of the incoming neutrinos. But other experiments such as SuperKamiokande (see Neutrino chapter) could. Its size, with 13,200,000 gallons of water looks quite impressive, doesn't it?

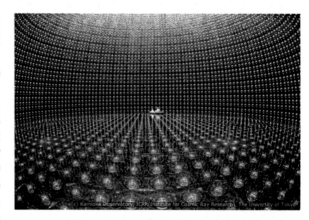

The SuperKamiokande experiment was indeed able to prove the theory of neutrino oscillation. Other experiments followed. And they did not only focus on the oscillation from electron- to muon-neutrinos. Also oscillation involving tau-neutrinos and all corresponding anti-neutrinos were set up and most of them are still measuring.

↗3:https://en.wikipedia.org/wiki/Neutrino_oscillation

Neutrino oscillation experiments are still very interesting, as the oscillations between the three types of neutrinos can be measured in order to determine the masses of the neutrinos. So far, people know that neutrinos do have a mass. Also, the differences between different neutrino masses are already known. But the absolute values are still missing. For this, different neutrino types need to be studied. Powerful neutrino sources are – next to the sun – nuclear power plants and particle accelerators. The latter type provides muon-neutrinos. Their oscillation into tau-neutrinos is studied by – among others – the OPERA experiment. It is located in the Gran Sasso mountain in the center of Italy and receives muon-neutrinos from the SPS particle accelerator. This accelerator is located about 450 miles away at CERN, close to Geneva in Switzerland. People are often surprised when they hear about the 450 miles that the neutrinos have to travel and ask for a very long pipe which should exist. But as we know that neutrinos barely interact, they can simply pass through the earth and rock between CERN and Gran Sasso.

Faster than Light?

The most impressive result of OPERA was in fact a measurement that was planned as a side-quest. Just for fun, the experimenters wanted to measure the speed of the neutrinos. By knowing the time when the neutrinos started their journey, as well as the distance that they have to travel until they are detected at Gran Sasso, you can immediately calculate their speed. No sooner said that

done! And the result, published in 2011, was… well… shocking. The measured speed was higher than the speed of light in vacuum. This is something that, assuming Einstein's theory of special relativity was true, is absolutely impossible. This meant serious trouble for a theory that seemed to be rock solid for decades. But no physics revolution started. Not only because physicists might not be the typical kind of people that start revolutions. It was also that many people did not believe in the result of the OPERA experiment. Not even all the members of OPERA did. But they tried their very best to find a mistake in any of their measurements and calculations. All their clocks would have to be synchronized with absolute precision at the level of nanoseconds. And they were. No mistake could be found. At least not until March 2012. It was a tiny detail that caused the wrong result. A cause so simple that you don't know if you should laugh or cry. Let's see if you can guess: If somebody tells you that his monitor does not work, what is the first question that you ask? Right: "Did you check the cables?". It was one of the fiber optic cables connected to a PC which was not completely in the socket. In contrast to a classic copper cable where you really need a close connection between the plug and the socket, the light passing through the fiber cable can cross a small gap. That's why you don't realize it immediately. But the gap causes reflexions, and therefore a shift in the time-of-flight of the optic signal. This leads to a mismeasurement of the neutrinos' speed. So remember: If you want to prove Einstein wrong, check your cables in advance.

Adventure in RADIATION THERAPY

DEEP IN THE TISSUES OF THE CELLULAR METROPOLIS, TUMOR-TROLL RELEASES HIS MUTATED EVIL ONTO INNOCENT CITIZENS...

IN A DINER NEARBY, OUR HEROINE PROTON GIRL IS HAVING LUNCH...

MY QUARKS ARE QUIVERING! ...SOMEONE NEEDS HELP!

KEEP THE CHANGE, BABY—DUTY CALLS!

WAIT! YOUR WEAPON!

YOU CALL IT A WEAPON...I CALL IT A TREATMENT.

SOON...

EAT THIS, TUMOR-TROLL!

AAAHG!! STREAMS OF IONIZING RADIATION...

MY KRYPTONITE!!

Radiation Therapy
Particles on a Mission against Evil

A loaded gun can do serious damage. But while being hit by a bullet can be pretty nasty, consider the fact that we are being hit by naturally occurring "bullets" every single day: ionizing Radiation. It is consisting of α-, β- and γ-rays (↗[1]) from radioactive decays (↗[2]) as well as X-rays. We are surrounded by this radiation coming from different sources. For example, high energetic particles reach us from outer space, hit the Earth at the upper atmosphere and cause cosmic particle showers (↗[3]). Also, bananas contain potassium which can decay radioactively as well as radon gas that comes out of Earth's surface. We call all of this type of radiation natural radiation.

Damaged Cells

Ionizing radiation ionizes atoms while traversing a material. This means they carry enough energy to take away electrons from atoms and turn them into charged ions. This fact is used in particle detectors (↗[4]) to detect such radiation as the ions can be converted into an electric signal. If ionizing radiation does not hit particle detectors but instead biological cells in our bodies, the created ions can cause chemical bonds to break or unwanted reactions to occur. What happens after damage occurs depends on the type of damage and dose of the radiation hitting your cells. The dose measures the energy deposited by the radiation per mass. If the dose is not too high – in particular the dose per time – our body has repair mechanisms to fix it.

If the DNA of a cell is damaged, a cell can – depending on the severity of the damage and the dose of the radiation – fix the DNA, mutate, or die. If the DNA is fixed, everything is fine. If the cell dies, it is also not too bad: millions of cells die and are being replaced within our body every minute. If the DNA is damaged and poorly fixed, in contains wrong modified construction plans and instructions for a cell's behavior. We call this a "mutation". In very few cases, mutations lead to changes which are not necessary bad. During the evolution of human beings, this happened numerous times. But most mutations will lead to degradations and in a worst-case scenario a mutating cell will seed a tumor, resulting in cancer.

People became aware of the damaging character of ionizing radiation at the beginning of the 20th century. The use of X-rays got quite common in the field of medical diagnosis as they allowed to get an insight to the human body. However, people realized that too high doses of X-ray radiation had negative consequences: people lost their hair and their skin got burned. Nowadays we know that we should protect ourselves against any kind of unwanted ionizing radiation.

But can we say that ionzing radiation is all evil, and no good? Consider the fact that we can use our ionizing particle guns to kill cells in our bodies that are less desirable, like tumor cells. Can we do that?

↗[1]: *"Alpha, Beta and Gamma Rays" on page 171*
↗[2]: *"Radioactive Decay" on page 167*
↗[3]: *"Cosmic Radiation" on page 257*
↗[4]: *"Particle Detectors" on page 253*

Yes, we can! Bad cells that we definitely not want are tumor cells. By radiating them with doses which are high enough to avoid any repair, you can get rid of them. This way of medical treatment is known as radiation therapy and can be used as an alternative to a surgical removal.

Using Radiation against Evil

Early attempts of medical treatment via radiation started with X-rays. In the 1950s, they were superseded by artificial radioactive γ-ray sources made of Cobalt or Cesium, which were produced in nuclear reactors. These sources provided larger energies compared to X-rays and allowed a deeper penetration into human tissue. In the 1970s, linear electron accelerators (↗5) allowed to produce very high energetic electrons. Patients could then be radiated with either the electrons directly or with photons to which the electrons were converted. Such linear accelerators are still used nowadays. There is one thing you have to be careful with: the cells which are supposed to be irradiated are surrounded by cells you don't want to damage. But the amount of energy that is deposited by the electrons and photons in the tissue decreases exponentially with the depth. This means, most of the energy is deposited at the surface. But the cells we would like to radiate lie at a certain depth. We know this problem when we transfer heat radiation to a sausage that we would like to grill. If we leave the sausage on the grill it will soon get burned at the surface but still be cold in the center. Maxwell knows the solution: rotating the sausages! During radiation therapy it is not the patient that is rotating, but the radiation source instead (more comfortable for the patient). This minimizes the radiation of healthy tissue at the outside but still deposits some energy at the center for each angle of radiation which adds up for a full rotation. This is practical for many purposes, but not all. Imagine a tumor deep inside your brain. A brain is too sensitive to radiate it from all sides. Often, surgeries are also no alternative. So what to do? For a long time there has been no answer. But then physicists and physicians found a solution. It is based on the different way that light and heavy ionizing particles deposit energy in matter.

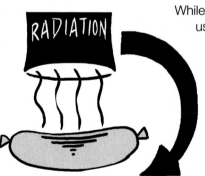

While for electrons and photons – used for standard radiation therapy – the amount of deposited energy is largest at the surface and then decreases, the situation is different for heavier particles, like protons: those release the major part of their energy not at the surface, but deeper inside the tissue. This point is called Bragg peak.

↗5: "Particle Accelerators" on page 249

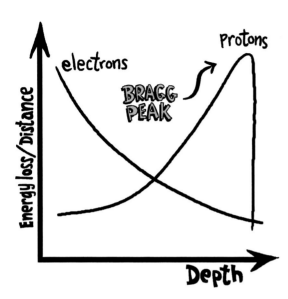

When physicists have to stop the beam of the LHC accelerator they shoot it onto a carbon cylinder which is 7 meters long and surrounded by 700 tons of iron and concrete. While for a classical bullet the hole would be at the surface of the block, a "proton beam bullet" would get stuck deep inside and leave its hole there. The depth at which a proton beam gets stuck depends on the position of the Bragg peak which itself depends on the proton energy. Now think back about the tumor deep inside the brain. By using a proton or ion beam and carefully calculating the position of the Bragg peak, you can radiate a brain without damaging it, except the tumor region deep inside.

Particle Accelerators in Hospitals

And in fact, this is what it is done at many hospitals! This field of radiation therapy is also known as particle therapy. It is quite expensive and complicated, but for many patients it is the only way to be treated. You do need your own proton accelerator, but fortunately not such a big one as the LHC with its 27 km circumference, as a human brain is not as hard as 700 tons of lead and concrete. In contrast to electron and photon irradiation, protons and ions deposit their energy in a well-defined region. By knowing the exact shape and location of a tumor, you can scan a tumor in the plane transverse to the beam by deflecting it with electromagnets. This same procedure is applied to an electron beam in the old (non-flat) TV screens. Even scanning the depth position is possible by varying the energy of the beam. This way you get a full 3D model which you radiate without much damage for the surrounding tissue.

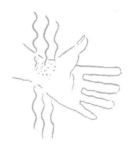

More and more clinics build their own particle accelerators to treat patients with electrons, photons, protons and ions. This is an example where elementary particle research, which usually has no purpose except a gain of knowledge about the structure of matter, has lead to a medical application. Physicists, biologists and physicians still keep researching in this field. A method which is currently under investigation (see the link to the ACE experiment, ↗[6]) is the radiation with anti-protons. Next to the Bragg peak advantage of protons you have the effect of annihilation of matter and antimatter where additional energy is released. This increases the damage (to bad cells) but makes the beam harder to control. Let us see what the future brings us!

↗[6]: http://home.web.cern.ch/about/experiments/ace

V – Beyond the Boundaries of Our Knowledge

"Physics is an amazing science – it tries to understand nothing less than to figure out the fundamental principles of our universe. How do things function, at a fundamental level? Physics has set out to answer this question. And we have come quite some way: With our knowledge of cosmology, particle physics and quantum mechanics, we have a pretty good grasp on the basic rules that govern the behavior of space, time and matter. Some phenomena might still be complicated (weather forecast, for instance), but the basic rules by which they function, are very well understood.

But we are not done yet! There are still some gaps in our knowledge of fundamental physics. Be it the unification of quantum theory and relativity, whether one can construct worm holes as shortcuts between different paths of the universe, the question of what happened before the Big Bang or the interpretation of the weird rules of quantum mechanics – there are several things we do not really know. Maybe, at some point in the future, the topics in the following, last part of this book will be an old hat. But for now, they are some of the most intriguing puzzles for the physicists of our time."

Exotic Matter
Different from Everything We Know

"Matter" – with this word, physicists usually mean everything that we are made out of, and even every kind of energy (¹) or quantum field. Basically, everything that is not space and time (although it seems probably that the distinction between space, time and matter wasn't always as clear as it is today, ↗²).

In particular, electromagnetic waves are also considered "matter", even though the elementary particles, the photons, which make up such a wave don't have any mass (↗³). Also antimatter (↗⁴) is considered "matter" in this context, which must be quite confusing, admittedly.

The universe is full of matter – it doesn't matter (no pun intended) if we know all of its properties precisely or not: dark matter (↗⁵), for instance, is quite mysterious to us, but most likely still ordinary matter. If we should find supersymmetric (↗⁶) particles one day, they would be considered matter. (By the way: as of the time of writing this book, the smart money seems to be on us not finding any, ever. Nobody has been able to observe supersymmetric particles at the LHC yet, despite a great effort to find some.)

All of these different kinds of matter, as diverse and individual they might be, share some common features. Most notably, they all have a mass which is greater than zero (or rather, which is not smaller than zero, see the photons, for instance).

There is only one other instance in this book where we have talked about particles without non-negative mass: particles with imaginary mass, called tachyons (↗⁷). But they most likely do not exist – they would either cause the whole universe to collapse instantly, or they would decay and fill the whole universe with a constant quantum field, of a certain value, and the quantum fluctuations around that value would again behave like normal matter (this is precisely what happens with the Higgs field, ↗⁸).

Matter: Always Positive Energy?

But here we would like to talk about another form of exotic matter, that is, particles with negative mass, or, equivalently, negative energy. Interestingly, although that sounds really weird, it is not actually forbidden by any of the fundamental equations of either general relativity or quantum physics. Most physical laws are about the exchange of energy. So

↗¹: *"E=mc²" on page 237*
↗²: *"Timeline of Our Universe" on page 100*
↗³: *"Light" on page 7*
↗⁴: *"Antimatter" on page 217*

↗⁵: *"Dark Matter" on page 133*
↗⁶: *"Supersymmetry" on page 299*
↗⁷: *"Tachyons" on page 291*
↗⁸: *"The Higgs Mechanism" on page 241*

energy gets increased at one point, and decreased at another point by the same amount. Allowing negative energy values would pose no problem for that (unlike the imaginary values for tachyons) – energy conservation (↗9) would still be satisfied everywhere, even if the energy would be negative at some places.

The equations of general relativity state that "the energy density of matter curves space and time". Nothing in this statement says anything about whether there is a minus sign in front of the energy density or not. If there were, the corresponding curvature of space and time would look quite weird, admittedly, but so what? Solutions to the equations with negative energy would include warp bubbles, allowing faster-than light travel (↗10), or wormholes, connecting different regions (or times) in the universe by shortcuts (↗11). It sounds like we'd need exotic matter to make space travel viable (or, at the very least, not quite as painstakingly slow).

Again, nothing like this is explicitly forbidden by the equations – it's just that, to the day we wrote this book, nobody has found any form of such exotic matter yet.

Lowering the Energy of the Vacuum: The Casimir Effect

There are, however, a few tricks that can be applied to create regions in space with net negative energy density. The most famous of it is the so-called Casimir effect: Place two large metal plates very close to each other, so that a narrow gap between the two remains. Now, as you may remember, because of the quantum nature of matter, empty space is not really empty – rather, it is brimming with lots and lots of particles which are created out of nothing, and vanish again fractions of a second later. That happens everywhere, all the time. Which energy the particles

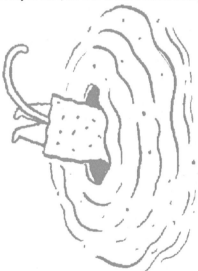

have which are created, is completely random with this process, all energies do appear.

But between the two plates however, there is only a very narrow space for the particles. Since every particle is also a wave (↗12), and the wave length is larger, the lower the energy, something curious happens: In the space between the Casimir plates, particles

---

↗9: "Conservation Laws" on page 51
↗10: "Warp Drive" on page 295
↗11: "Wormholes" on page 287
↗12: "Wave-Particle Duality" on page 143

appear out of nothing and go back to nothing, but only those above a certain energy! For particles with very low energies, there is not enough space – if the wavelengths are larger than the gap between the particles, the wave cannot form. So effectively, between the two plates there are fewer quantum fluctuations happening than outside of the plates, because some particles cannot appear. Between the plates, the average energy density is therefore lower than outside (which is just vacuum).

Now you might say that this is cheating: it is not that there is actual negative energy density, it is just that the vacuum has a very, very high energy density, and between the two plates there is still a high energy density, but just a bit lower than outside. But, as we have already said, physical laws are just about energy differences. If we would just increase the energy of everything in the universe by the same amount, physics would not change! And if we declare the energy of the vacuum to zero, then the Casimir effect actually generates a region with negative energy!

Exotic Matter – Mysterious, but Not Impossible!

It should be noted that this effect has actually been measured: in the laboratory, it could be observed that there is a tiny force between the plates. In fact, it is the vacuum itself that pushes the two plates to-

gether – a sure sign that there is less energy inside than outside of the two plates!

The effect is far too tiny to create a stable worm hole, sadly. Even if it weren't, the space-ships that would want to travel through, would have to be really, really slim: For a significant force to exist between the two Casimir plates, they would need to be only a few hundred atoms apart. Talk about unrealistic body expectations for astronauts!

Anyway, apart from these quantum tricks, we do not know if there is actually, anywhere in the universe, some sort of exotic matter which has negative energy density. Apart from the very strange physical behavior of this type of matter (it would have negative pressure as well, for instance), it is quite hard to predict how it would actually interact with other kinds of matter. Could we hold a piece of it in our hands at all? Hard to say. If not, then we could probably not even see it – it wouldn't interact with the electromagnetic field. In that respect, it would be very much like dark matter, although with basically the opposite gravitational effect.

As strange as it sounds, we should not disregard the possibility that exotic matter really exists. The universe has, in the past, managed to surprise and amaze us again and again, and it will certainly continue to do so. Maybe, one distant day in the future, we'll pop to the supermarket in the Andromeda galaxy via a wormhole held open by exotic matter.

Before the Big Bang
A Bounce?

It has been known for nearly a hundred years that the universe is expanding. First observed and made public by Erwin Hubble, the result was later understood theoretically by Albert Einstein: unless the universe is completely devoid of any matter and radiation (which ours certainly is not, otherwise there'd be no one to read these words right now), the theory of general relativity dictates that it has to either expand or contract.

The equations also, unmistakably, tell us another fact: the universe did not expand forever – in the past, there must have been a point in time where it all began. At the moment, our most precise measurement put this moment – when the whole universe was compressed to one minuscule point of incredible density and heat – at about 13.8 billion years in the past (↗[1]).

13.8 Billion Years Ago: The Big Bang

Now one could ask: what happened 13.9 billion years ago? Unfortunately, we do not even know if this question makes any sense.

If you just look at the equations of general relativity, then that question really cannot be answered. All of the physical quantities, such as temperature, density, space-time curvature, tend to infinity, the closer you come to the point which is referred to as the Big Bang. There is no time, and no universe earlier than that. Asking what was before the Big Bang is like asking "What is north of the North Pole?" There is nothing north of the North Pole, because it is,

by definition, the point furthest north on the planet (of course you can ask "what is above the North Pole?", but that is simply not the same question). If the shape of the universe is completely described by general relativity, then there is not "before the Big Bang", for the very same reason.

Make no mistake: as weird as that sounds, it might just be something we have to live with. For the human mind, which has evolved over millions of years to think about the world in certain categories, that might be a particularly hard pill to swallow. We are just so used to think in terms of cause and effect, that the thought that there was a first point in time, seems unbearable. But if the physics of such a situation is sound, and agrees with every observation we can make about the universe, it might be a notion we just have to make our peace with, even if our minds are not made to grasp it. Why should nature be kind enough to behave in a way, so that some descendants of hairy mammals on a small planet in the outskirts of a minor galaxy have an easy time picturing it?

General Relativity Does
Not Tell the Whole Story

Now, we have several reasons to believe that general relativity does not tell us the whole story, however. One of them is that the equations do not take any quantum effects into account. A unification of quantum theory and general relativity (↗[2]) should be a much more realistic description of what was happening 13.8 billion years ago. The reason why

↗[1]: *"The Big Bang" on page 97*
↗[2]: *"Quantum Gravity" on page 279*

physicists think that, is that quantum theory tends to avoid singularities, which are rampant in classical physics.

Here is an analogy: Think, for instance of the classical description of a proton and an electron, as a positively, and a negatively charged point particle. It is easy to compute the force between them with classical electromagnetism: The force between the two gets stronger the closer the two approach each other. Now think of the two just sitting next to each other. What will happen? Of course, the electron will start falling towards the proton. Now comes the point: after a few seconds, when the electron hits the proton, the force between the two becomes infinite! When that happens, the classical equations describing the motion of electron and proton break down, and cannot be used to calculate what happens beyond that time. They cannot deal with infinite numbers!

Of course, that is not what happens in real life, because electrons and protons are not ideal point particles. That is a good approximation when the two are far away from each other, but as soon as the two come very close to each other the quantum properties of the two particles become important. Quantum mechanics, however, tells us that there is a minimum distance that the electron can be from the proton,

because of Heisenberg's uncertainty principle. A proton and an electron at that distance are what's called a hydrogen atom, by the way!

Quantum Gravity to the Rescue?

Many physicist assume that the way physical quantities become infinite at the Big Bang in general relativity, is the same way in which they become infinite in our analogy example with the proton and the electron: it's not what is really happening – taking quantum theory into account will give a more complete picture.

There is even more reason to believe that there was no true singularity at the Big Bang: there are quantum gravity models (↗²), in which there is some kind of Heisenberg uncertainty relation for distances – so they suggest there is a minimum, non-vanishing length (that is about the Planck length, roughly 10^{-35} m). But if there is a minimal length, there also is a minimal volume. And because "density" is "mass divided by volume", that also means that there is a maximal, physically possible density! Very, very large (the so-called "Planck density", about as many suns as there are atoms in the universe, in a gallon), but not infinite. Now, if the physical quantities are just very, very large, but not infinite at the Big Bang, one might actually have a chance of something happening before it!

The Big Bounce: Connecting Two Universes

At the moment, quantum gravity theory is still in the stages of being developed. However, a few years ago, researchers at Penn State University were able to use the equations of their latest quantum gravity

universe starts to expand – and this is the amazing thing – exactly as if a Big Bang had just happened.

Although many of the details are still unclear, this suggests that, rather than a Big Bang, there was a "Big Bounce". Before the expansion of our universe, it could have been a contracting one.

So there might be a way to ask "What happened 13.9 billion years ago"? Unfortunately, not quite.

model, to compute what might happen during the Big Bang. They found something surprising: If one starts from a universe which is slowly contracting under its own weight, it will do that, until the density of the universe becomes about as high as the Planck density. Then, gravity suddenly becomes repulsive. The quantum fluctuations in the gravitational field become large, so that the universe does not collapse – rather, it undergoes a phase of… let's call it "quantum strangeness". It is crushed like a crumpled-up piece of paper, but, and that is important, it is not compressed to a point! After that phase, the

During the phase of quantum weirdness, which connects the Big Crunch of the old and the Big Bang of the new universe, space and time are really, very, very quantum. So much so, that it does not make any sense to ask how long that phase lasted. To do so, one would need a more or less normal flow of time – and time and space are so crumpled up during that phase that one cannot really say how long it took to go from contracting to expanding.

It looks like a Big Bounce scenario might be even weirder than the Big Bang one!

Quantum Gravity
Where Is the Quantum Theory of the Fourth Force?

At many other points in this book have we pointed out that there are four fundamental forces in nature that we know of: gravity ($\nearrow^1$), electromagnetism ($\nearrow^2$), the strong ($\nearrow^3$) and the weak interaction ($\nearrow^4$). The latter two only play a role in the quantum world of atoms and elementary particles, because their range is so incredibly short. The first two are the ones we also experience in our classical world, where they act between planets, chairs, magnets, cats, and so on. But all of these are built up from elementary particles, so the gravity and electromagnetic force between large (macroscopic) objects should be a consequence from the same force acting between small (microscopic) objects, right?

Gravitons: It's Not as Easy as It Looks!

For the electromagnetic force we know pretty well how this works. On a microscopic scale it is mediated by photons, which fly around between charged particles. Also, a large number of photons flying in the same direction form a macroscopic electromagnetic wave ($\nearrow^2$). But how about gravity? Well, we also know that in our classical world, there are gravitational waves ($\nearrow^1$). And physicists have postulated that there could be "quanta of the gravitational field", which have been dubbed gravitons. In the 70s of the last century, they tried to write down a quantum theory of gravitons along the same lines as the well-known and well-established quantum theory of photons. But unfortunately, they found out that this does not work. The formulas made no sense, and

delivered unphysical results. It appeared that gravity, as an interaction, did not fit into the framework of standard quantum theory.

Up to the present day, we still don't have a quantum theory for the gravitational interaction. But there are some good ideas how to proceed. Also, people have an idea why the initial attempts of constructing one failed: it treated gravitons pretty much the same way as photons – as little particles traveling through an unchangeable and fixed space. Space and time were a stage, and the particles moving in it were the actors. For the other three interactions this is a valid way of going about, but gravity is all about like space-time itself being an actor – masses curve and change space-time, which in itself tells particles where to move. Space-time and particles should be treated as co-actors in this play of the universe.

No Big Bang without Quantum Gravity

All right, so the initial attempts of constructing a quantum theory for the gravitational interaction failed – how about we make some experiments to see what is going on? You can measure single photons, should one not try to measure single gravitons, or whatever constituents of quantum gravity there are?

Well, here the problems with quantum gravity continue – it is incredibly hard to find phenomena in nature for which it is not already sufficient to use either

---

$\nearrow^1$: *"The Theory of General Relativity" on page 121*
$\nearrow^2$: *"Light" on page 7*
$\nearrow^3$: *"The Strong Interaction" on page 229*

$\nearrow^4$: *"The Weak Interaction" on page 233*

quantum physics or general relativity to understand it. This is because quantum theory is for the description of very small things, like elementary particles, and general relativity is for very heavy things, such as stars and galaxies. A situation in which both quantum and general relativistic effects are important would need to be very extreme indeed.

And make no mistake, there certainly are such physical phenomena: in the center of a black hole ($\nearrow^5$) the mass of an entire star is compressed into the space of less than an atom. Also, at the Big Bang ($\nearrow^6$), the mass of the whole universe was compressed into a tiny point, before the universe started to expand. To understand those two phenomena, one would need a theory of quantum gravity, because both quantum and relativistic physics play a role. But we don't have one, which is why we do not know precisely what happened at the Big Bang, or what goes on in the singularity of a black hole. And it's not like one could just go into the drug store and buy one to make measurements, unfortunately. Well, probably fortunately.

Space and Time Quantized: Planck Length and Planck Time

Even if we do not know what the quantum behavior of gravity is, there is a very good argument that one of the consequences should be that there is a smallest non-vanishing length. The argument is as follows: Assume you want to resolve a very small distance. For this you need, for instance, a photon with a wavelength which is at least as short as the distance one wants to look at. Now on the one hand Quantum Theory tells you that this means the photon needs to have a very large energy. General

relativity on the other hand tells you that if you concentrate enough energy into a small enough space, it will curve space and time into itself, forming a black hole ($\nearrow^5$). And there is no way of seeing or measuring anything behind the event horizon of a black hole.

A quick estimate reveals that this happens, whenever one tries to measure anything about as small as the Planck length of $l_{Planck} \approx 1.6 \cdot 10^{-35}$ m. In other words, there is no physical experiment which can resolve distances smaller than the Planck length. So this is really the smallest distance which makes any physical sense. Similarly, there should be a shortest physically meaningful duration: the Planck time of $t_{Planck} \approx 5.3 \cdot 10^{-44}$ s.

If you ever heard of Zeno's paradox of the Tortoise and Achilles, you know that this would solve a millennia-old mystery ($\nearrow^7$)!

A Viable Candidate: Loop Quantum Gravity

One attempt to merge gravity with the other three fundamental forces is pursued by String Theory. Although here space and time are still treated as a fixed stage, there is some hope that one can get around the problem of nonsensical formulas that the physicists in the 70s and 80s had. We will talk more about it in the respective chapter ($\nearrow^8$). Here, we would like to talk about another idea, which has found some traction in recent years: Loop Quantum Gravity.

One of the basic ideas in Loop Quantum Gravity is that space is not a continuum of points, but rather consists of quanta of space, also called nodes, which are connected to form a huge network-like

$\nearrow^5$: "Black Holes" on page 91
$\nearrow^6$: "The Big Bang" on page 97
$\nearrow^7$: http://en.wikipedia.org/wiki/Zeno%27s_paradoxes

$\nearrow^8$: "String Theory" on page 303

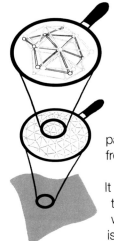

structure. A matter particle can then not be anywhere, but only sit on top of these nodes. An exchange boson, however, can only sit on the links connecting two nodes – because it describes the force a matter particle feels when hopping from one node to another.

It might be difficult to imagine that, rather than a "room in which things move", space is more like a "network along which things move". This network (also called "spin network", ↗[9]) should not be thought of as being somewhere in space, rather it is space itself! Just as the atoms, which you can see when you look at a table under a strong microscope, don't lie on the table surface – rather, they are the table!

Space-Time Foam Instead of Space-Time?

When the theoretical physicists began to calculate what the consequences of this idea were, they were quite surprised: the length of a link – in other words, the distance between two nodes – can not take arbitrary values, but is quantized. In other words, there is a smallest non-vanishing length, just as one would have expected from a quantum gravity theory! And it is indeed roughly (although not precisely) the Planck length.

This is very encouraging, but so far we have only described space – what about time? Shouldn't

the two be connected? Yes, indeed they are: the three-dimensional network can indeed change over time, sweeping out a four-dimensional structure called a space-time foam. The name comes from the fact that the one-dimensional links changing over time form two-dimensional faces in space-time, which are touching at various lines, giving the impression of soap films.

So space is discrete in this theory, but is time discrete as well? Yes it is: although it looks like a space-time foam would describe a continuous change of a network, the geometric properties such as lengths, areas and volumes only change in discrete steps. And what is time, other than change of physical properties? So because change happens in discrete steps, time also progresses in discrete steps.

The concepts of Loop Quantum Gravity (also explained in a 5 minute video at ↗[10]) are intriguing and look quite promising. Alas, at the point of writing this book, there have actually been very few actual solutions to its equations, because the mathematics is so very complicated. However, there are some very fascinating intermediate results about what happened at the Big Bang (↗[6]). So stay tuned!

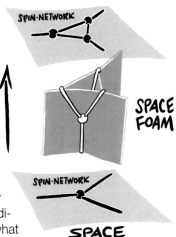

↗[9]: http://www.einstein-online.info/spotlights/spin_networks
↗[10]: https://youtu.be/_ShR3Zx0y2k

"Incredible. I wasn't aware planck stars had planets orbiting around them, let alone a spacefaring species."

Black Hole Evaporation
Planck Stars Instead of a Singularity?

Black holes are among the most mysterious and awe-inspiring phenomena in the universe (↗1). They consume everything around them, shredding matter into tiny bits and pieces, while freezing what's left, in time for all eternity.

The very core of a black hole seems to be a place which defies all known laws of physics: It warps space and time around itself so much, that it is hidden from the rest of the universe behind a barrier that only goes one-way: the event horizon. The center itself is a place with such strong space-time curvature, that all our common notions break down if we try to compute it – Einstein's equations just claim that it is infinite. And it is always growing heavier, sucking in more and more matter from its surrounding.

But is it really true? Can a black hole never die? Or do these ravishing behemoths have a finite lifespan, such as everything else in the universe?

Black hole Equations: Incomplete without Quantum Physics!

Everything that we know about black holes comes from the study of the equations published by Einstein in 1915 (↗2). If you follow Einstein's equations, the answer is clear: there is no way to unmake a black hole – everything you do in order to change it will always make it larger. But even Einstein remarked, already one year later in 1916, that his equations needed to be modified in order to take the effects of quantum physics into account. This feat has still not been accomplished completely though, to this day (↗3).

Even if you do not have a fully-fledged unification of the principles of space, time and the quantum, however, there are some statements that you can make about black holes. In 1974, one young physicist called Stephen Hawking, thought hard about it and wondered: if everything that crosses the event horizon is lost and is added to the total mass (↗4) of the black hole, is it possible to decrease its mass by sending in something with negative energy? At that time, he did not think about exotic matter (↗5), but about conservation laws in quantum physics (↗6), and vacuum fluctuations.

We know that, because of quantum physics, empty space is actually never completely empty – rather, the vacuum constantly has pairs of particles and antiparticles appearing out of nothing, and vanishing back to nothing again. For a short moment, energy conservation can be violated, and mass/energy can be generated. The larger this violation, however, the shorter it can be sustained. In the long run, energy conservation must be satisfied.

Hawking and His Radiation

Now what if, Hawking mused, a particle-antiparticle pair was created very close to the event horizon,

↗1: *"Black Holes" on page 91*
↗2: *"The Theory of General Relativity" on page 121*
↗3: *"Quantum Gravity" on page 279*
↗4: *"E=mc² on page 237*
↗5: *"Exotic Matter" on page 271*
↗6: *"Conservation Laws" on page 51*

just outside of it? And what if, furthermore, their paths would lead one of the two away from it, while leading the other across the event horizon, into the black hole? According to Hawking's calculation, the particle that escapes the black hole carries a certain amount of energy, and because energy conservation has to be obeyed in the long run, that energy has to come from somewhere. And it does: it has been taken from the black hole!

The actual calculation is not that difficult (as with all revolutionary ideas, a couple of decades later everybody looks at it and says "It's obvious!"), but the physical interpretation is not that easy. Another way of interpreting what happens, is that the two virtual particles, which are allowed to break energy conservation, are turned into real particles after some time of separation. But real particles have to satisfy the law of energy conservation, and since the escaping particle has positive energy, the captured particle has to have negative energy, effectively reducing the mass of the black hole after falling into it.

Yet another – equivalent – way of interpreting what happens is that the energy that

is needed to create the particle-antiparticle-pair is taken from the gravitational field of the black hole. If one escapes and the other does not, only half of that borrowed energy is returned to it, so the gravitational energy – and hence the mass – of the black hole suffers some small loss.

Do Black Holes Evaporate?

Whatever the precise way you'd like to interpret what comes out of the calculation, it is firmly believed nowadays, that there exists a way for black holes to lose energy. In other words – they are not completely black! Rather, the particles flying away from them look like some kind of heat radiation from far away. This is the so-called Hawking radiation. It makes it look like black holes have a temperature higher than absolute zero!

But what's even more interesting is that, by gradually losing mass, black holes can eventually evaporate into nothingness. Well, at least in principle that is. A sun-sized black hole has a temperature of about a few nanokelvin – that is much less than the temperature of the cosmic microwave background

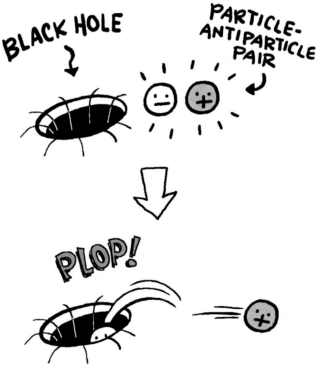

(↗7). So the afterglow of the Big Bang, which fills the whole universe, is replenishing the loss of mass faster than the Hawking radiation can take it away. However, if one were to encounter, somewhere in the universe, a black hole which weighed only a few kilograms, then it would radiate much faster, evaporating in a couple of weeks! Such a black hole could not have been created via the usual mechanisms in the universe, though: black holes that come into being as result of the death of stars have to have at least a couple of solar masses (↗8).

Actually, before the particle accelerator LHC was turned on in 2010, some people were afraid that its energies would be large enough to create tiny black holes, which then would grow and eventually destroy the whole Earth! Now, as of 2016, when this book is being written, the world is still here. But even if some microscopic black holes had been created during some of the runs of the LHC, chances are high that they would have been so light that they had evaporated within fractions of a second!

All of Hawking's calculations, however, are only good for stellar-sized black holes. For the really small ones, the evaporation process is not understood at all. One can estimate that the Hawking radiation becomes stronger, the lighter the black hole is, but what would happen in its final stages, in particular at the moment when the last bit evaporates away, is something we have no idea of right now.

There are some (highly speculative) ideas, coming from quantum gravity researchers, though.

Planck Stars and White Holes

One is that, instead of a singularity, there is a so-called Planck star at the center of the black hole. This is supposed to be a very small region, in which space and time are highly quantum in nature. Curvature is very large, but not infinite. The gravitational contraction is being counteracted by quantum mechanical effects themselves, most notably the Heisenberg uncertainty principle. This would keep it stable.

A shrinking black hole (which already would need to be quite small, about the mass of the Earth or lighter), would have an ever decreasing event horizon, while the Planck star in the interior would grow in size, the more mass falls into the black hole. When the two would meet each other, the whole matter trapped inside would be released spontaneously, in an enormous explosion, very similar to what the equations of general relativity describe as a white hole. At the moment this is highly speculative, but researchers from the Centre de Physique Théorique in Marseille claim that there are actually signals that have been measured by astronomers, which look very similar to what a white hole would look like. Whether there actually are any white holes in the universe or not is unclear at this point, but there are some exciting possibilities. Maybe black holes are not immortal after all.

↗7: *"The Cosmic Microwave Background" on page 105*
↗8: *"Supernovae" on page 83*

Wormholes
Shortcuts through Space and Time

Einstein's theory of relativity ([1,2]) states that it is impossible to accelerate anything from rest to some velocity faster than light. It seems depressing – the dream of whizzing around in space ships to reach the far corners of our galaxy in just days will probably never come true.

The long way

Well, at least not if the space ships travel in conventional ways. Incidentally, the theory of general relativity allows for some unconventional ways of travel. With these one might be able to reach far away stars and planets in a short time, turning the science fiction dream of space travel from a hopeless fantasy to a remote possibility.

munch munch munch

One of these possibilities is the wormhole. A wormhole is like a tunnel, connecting two parts of the universe. Even though these two regions can be light years apart, the tunnel itself could only be a few yards in length. A space ship could fly through the tunnel at a leisurely pace, crossing a region of space it would have needed thousands of years to go to conventionally.

The reason that wormholes are allowed by Einstein's relativity theory, is that its existence rests on the fact that space-time can be curved ([3]). Imagine a sheet of paper, and an ant, scurrying around on it. If the ant would like to go from one edge of the sheet of paper to the other one, it could just slowly crawl along the whole width of the paper. That would take quite some time, but imagine that the ant could fold the sheet of paper somehow, so that the far away spot were to lie exactly on top of the spot where the ant is now. All the ant would have to do then is to gnaw a little hole into the paper, and make its way through it. It would emerge on the other side, saving a lot of time! By the way: any other inhabitants of the sheet of paper would not realize that their whole world were bent – the intrinsic curvature of the sheet of paper would not change during such a process ([3]).

The Smart way

All right, relativity theory does not forbid the existence of wormholes. But how would they actually look like, and could we build one?

The Einstein–Rosen Bridge: Black and White Holes

The first realization that it could be possible to have tunnels between different regions of space is actually quite old: The so-called Einstein–Rosen bridge is a theoretical possibility of a connection between two regions of space (or between universes). It is a one-way street: its entrance looks exactly like a black hole ([4]). Just that a particle falling into it

[1]: *"Relative Space and Time" on page 117*
[2]: *"The Theory of General Relativity" on page 121*
[3]: *"Curved Space Time" on page 125*
[4]: *"Black Holes" on page 91*

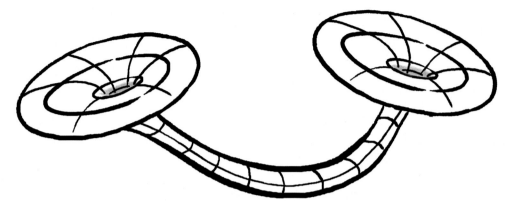

would, instead of reaching the singularity, emerge in another part of space, exiting the wormhole through something that is called a white hole. (Just as nothing can ever escape a black hole's event horizon, nothing can ever approach a white hole closer than a certain distance). But not only would the gravitational forces be extremely strong in its interior (space-ships would very likely be crushed instantly), it is also extremely unstable! The reason for this is that white holes are so very fragile (which is why one reads so little about them – whenever one would exist, it would collapse immediately, ↗5). The time for such a worm hole to collapse is much, much shorter than it would take a particle to cross from one end to the other.

The Morris–Thorne Wormhole: a Viable Alternative?

End of the twentieth century, the physicist Kip Thorne, together with his student Mike Morris, discovered that there is a way to have wormholes that are not only stable, but could also be traversed in both directions. The caveat with this wormhole is that it needs exotic matter (↗6) in its interior to keep it open.

Here is the crux with all types of wormholes: at some part of their interior, it seems that there needs to be a region in which gravity becomes repulsive, rather than attractive. The reason for this is the following: assume that lots of light rays converge on the entrance of the wormhole, all from different directions. When they pass through the worm hole and emerge on the other side, they need to diverge, traveling off in different directions. So, at some point within the worm hole, the converging light rays needed to be repelled by each other. In other words, space-time in a worm hole needs to be curved in such a way that it looks, effectively, like gravitational repulsion (The exact argument is a bit more complicated, because light rays refracted by e.g. the gravitational field of a star can also send converging light rays on diverging paths, without the star having repulsive gravity anywhere. But such a star has a focal point, because it acts like a gravitational lens (↗7), which a wormhole has not. That together with the diverging light rays means that gravity needs to be repulsive).

Exotic Matter and Repulsive Gravity

However, all known forms of matter that we know curve space-time in a way which leads to gravita-

↗5: *"Black Hole Evaporation" on page 283*
↗6: *"Exotic Matter" on page 271*
↗7: *"Gravitational Lensing" on page 129*

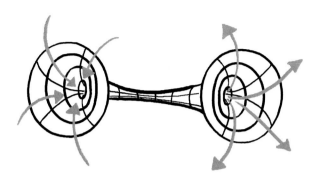

Converging light rays   ## Diverging light rays

tional attraction (even antimatter! ⬀8). This is because all types of matter that we have encountered so far have positive mass (or at least, positive energy, ⬀9). The equations of general relativity are straightforward in that regard: if you want to have gravitational repulsion, you need regions with negative energy density. Now, there is nothing which tells you that there cannot be matter with negative mass. In fact, there are some ways to lower the energy density below zero in some regions by quantum effects (this happens in a way which is useless for building wormholes, unfortunately). It's just that we have never encountered something like it, which is way physicists have coined this – hypothetical – form of matter "exotic".

Either way, if we had such exotic matter, it would theoretically be possible to construct wormholes between different regions of space. There are even versions in which a space-ship traversing the wormhole could stay clear of the region with exotic matter.

Time Travel with Wormholes: Just a Space Ride Away!

By the way: if you have a wormhole which connects to far away regions of space, then you can use this to build a time-machine in no time (no pun intended)! All you need is the wormhole version of the twin paradox (⬀1): Imagine you have a wormhole, the two entrances of which are quite close to each other. If you enter one side, you appear immediately a few yards away on the other side. Now all you have to do is move one entrance of the wormhole around – put it in a space-ship, which flies at nearly the speed of light, but by conventional propulsion. Drive the space ship far away and come back a few hours later. For the one entrance of the wormhole, which was moved on the space ship, just a few hours will have passed. But for the wormhole entrance which stayed behind, a few years will have passed because of relativistic time dilation.

Now the two wormholes connect not only different regions in space, but also different times: If you enter one side, you will vanish and emerge a few years later on the other side. However, this being a time machine, you could emerge on the one side of the wormhole a few years before you enter the other one. Of course, this is predestined for all different kinds of time travel paradoxes. Maybe it is good that we have not yet found a way to build a wormhole, after all.

⬀8: "Antimatter" on page 217
⬀9: "E=mc2" on page 237

Tachyons
Actually Faster than Light

It is often claimed that nothing can go faster than light. But why is that the case? Well, didn't Einstein say so?

Actually, Einstein's theory of relativity says something slightly different: it says that E=mc², where m is called the accelerated mass (↗1). It is not the actual mass of a particle, but equal to $m = γ\,m_0$, the product of the actual mass (also called rest mass), and the gamma-factor γ. It is this factor which causes problems when trying to accelerate any particle beyond the speed of light.

Infinite Energy: Reaching the Light Barrier

You see, the value of this gamma-factor depends on how fast the particle actually moves. Think of a particle moving through space. Its velocity is a fraction β of the speed of light. If β is zero, then its velocity is zero, and the particle is at rest. If β is one, then it travels at precisely the speed of light. In a formulaic expression, the value of the γ-factor depends only on β, and it is

$$\gamma = \frac{1}{\sqrt{1 - \beta^2}}$$

If you have some familiarity with formulas, you will see what happens when β goes from zero to one: The gamma factor, and hence the energy, increases ever more and more. In other words: to have a particle accelerate from rest mass to the speed of light, you actually need an infinite amount of energy! So no can do!

There is a well-known exception for photons (and other particles that have no mass, if there are any – we don't know of any, ↗2): They always travel at the speed of light, and can never be slowed down or accelerated to any other speed.

Tachyons: Always Faster than Light

But, theoretically, there is also another possibility: if the speed of a particle were larger than the speed of light, then the value of β would be larger than one. That would mean that under the square root in our formula for γ, there would be a negative number. And the square root of a negative number is an imaginary number! This weird particle could, however, still have a well-defined energy, if at the same time its mass was imaginary, too. The imaginary unit of the mass and the square root would cancel each other out, and everything would be fine, at least from the side of the mathematics.

Such a particle, with an imaginary mass and which travels faster than light is called a tachyon. It is not, strictly, forbidden in Einstein's theory of special relativity, but let us be clear right from the start: nobody has ever seen such a guy!

The Tachyonic Antitelephone: Calling Yourself in the Past

A tachyon, as described by Einstein's theory, would have some really strange properties. The first and foremost would be that it would travel faster than light, of course. That, in itself, would bring all kinds

↗1: "E=mc²" on page 237
↗2: "The Higgs Mechanism" on page 241

of problems with it (↗³). It would violate the principle of causality, which is a central concept in both relativity and quantum theory, and which guarantees that cause and effect happen one after the other.

With a tachyon, however, you could break this. You could not directly travel through time, but so something nearly as bad: you could transmit information to the past. That is actually really easy: Just send a message encoded in a beam of tachyons to a colleague, who is moving away from you very quickly. All he would have to do is to send the message back to you with the same type of tachyons. A quick calculation in special relativity reveals that the signal would arrive at your lab before you had sent it away.

The reason for this is that the event of you sending the message and your friend receiving it are what is called space-like separated. For two such events, one cannot say that one has to definitely happen before the other – that depends on how an observer, being asked to judge that, moves.

Knowing this, the other strange properties of tachyons seem tame in comparison: it would be impossible to slow a tachyon down to the speed of light. In fact, it would have less energy, the faster it traveled. It would be very easy to make a tachyon as fast as you wish, without any boundary. But to slow it down to the speed of light, you'd need infinitely much energy.

So the speed of light is the ultimate barrier – impossible to cross from either side!

Quantum Effects: Causality Is Restored …

Now, all of these considerations do not take quantum theory into account. If you were to pay attention to what quantum effects you would get with such particles, you'd realize that the situation changes dramatically!

First of all, in quantum theory, particles are just wobbles in a quantum field (very loosely speaking). Properties like the mass of the particle translate directly to the stiffness of the field, and how easy it is to make it wobble!

If one does the calculations carefully, one finds that the quantum field excitations with imaginary mass do not travel faster than light. A ripple in the quantum field propagates in the same way as if the field would have an ordinary, real mass.

… but the Universe Implodes Instantly!

However, something different, totally catastrophic, would happen. The energy of a ripple in a quantum

↗³: *"Relative Space and Time" on page 117*

field depends on the square of the mass. Now if that becomes negative, because the mass of the field is imaginary, then one can gain energy by creating a tachyon out of nothing! And one important rule in particle physics is that all that can happen, will happen – and the more energy is released by the process, the more likely it is happening.

So if there was a quantum field with imaginary mass, and nothing else to stop it, a cascade of tachyons would appear out of nothing – and more, and more, releasing more and more energy, in a flood that would never stop. The whole universe would be immediately filled with more and more tachyons. This process has been termed tachyon condensation, and it is something very undesirable – with so many particles, the universe would collapse into itself in an instant (↗4). This is the reason why there are no (free) tachyons.

> DO I EVEN EXIST?

Incidentally, the original formulation of string theory (↗5), which was formulated in 26 dimensions, had some ways of the strings to wobble, such that the resulting particle would have imaginary mass, and therefore this whole version of string theory was tossed quickly – tachyon condensation was something that no theory should predict, if it were respectable!

The Uncondensed Higgs Particle Is a Tachyon!

However, there is one known exception to this: the Higgs particle (↗2)! Technically, the Higgs particle is a tachyon, since in the formulation of the quantum field its mass appears as imaginary. But why did the universe not implode yet? Well, in the formulas for the Higgs field, there is another term beyond the mass term – if a certain density of Higgs particle has been reached, it is energetically really inconvenient to generate any more. But up to this maximum, the tachyon condensation has happened with the Higgs particle! This is why the Higgs field fills the whole universe, giving all particles matter: the universe is filled with a lots and lots of Higgs particles, having condensed out of the vacuum in the beginning of the universe.

What people nowadays call the "mass of the Higgs particle" by the way, is not the imaginary mass term in the formulas. Rather, the (positive) mass of the Higgs that has been found to be around 126 GeV describes how difficult it is to generate new Higgs particles, now that the universe is already filled up by them!

↗4: *"The Theory of General Relativity"* on page 121
↗5: *"String Theory"* on page 303

Warp Drive
Surfing on a Space Time Wave

To travel faster than light – it seems that there is no way to reach far away star systems, unless one finds a way to circumvent the central statement of both special and general relativity ([1,2]): nothing can travel faster than light.

Space and Time Can Expand Faster than Light

This statement actually needs to be worded more carefully: nothing which is at rest can be accelerated to reach the speed of light in a finite time. But one could use shortcuts to reach far away regions of space ([3]). Are there other ways around the speed of light?

There is another subtlety about this: The maximum speed is only valid for matter moving in space – not for space itself. This means that, for instance, far away galaxies effectively move away from us with a speed several hundred times the speed of light. But it is not the galaxies that are moving here. Rather, the universe between us and them expands very fast. And there is no maximal rate of expansion of the universe, at least as far as we know!

This inspired physicist Miguel Alcubierre, at the end of the last century, to search for a way to use expanding and contracting space to move a spaceship faster than light. Well, as he admitted later, the TV series Star Trek was also a major inspiration.

And he succeeded: Alcubierre found a theoretical possibility for faster-than-light travel: it is possible to move a space ship from one point in the universe to the other faster than light – by not moving the vessel at all, but by contracting the space in front of it, and expanding space behind it.

The Alcubierre Proposal: a Real Warp Drive?

The region in between, where the space ship resides, is called the warp bubble. In it, the ship is completely at rest, and feels no acceleration in any direction. So the astronauts have a quite comfy ride, while their ship surfs on a wave of curved space-time through the universe. Within the surface of the warp bubble, however, there are tremendous tidal forces, so everything and everyone trying to cross the border between the inside and the outside would be torn to shreds in an instant.

Even worse, the space travelers will, unfortunately, not be able to enjoy the view during their trip. The inside and the outside of the warp bubble are what's called causally disconnected. This means that it is impossible to send a light signal from the inside of the bubble to the outside, or the other way round. These two regions of the universe are separated by a horizon, similar to the event horizon of a black hole ([4]).

[1]: *"Relative Space and Time" on page 117*
[2]: *"The Theory of General Relativity" on page 121*
[3]: *"Wormholes" on page 287*

[4]: *"Black Holes" on page 91*

So what would it take to build a ship with warp drive? Could a vessel just generate a warp bubble on its own, just as in the TV series?

Again with the Exotic Matter!

Well, as you have might have guessed, it is not that easy. The way space and time are curved in the warp bubble is quite special, actually. The curvature is such that no ordinary matter could ever be used to cause it. Rather, one would have to have large amounts of exotic matter ([5]), which would need to have negative mass – or negative energy density, that would work equally well ([6]). Unfortunately, although theoretically possible, nobody has ever encountered any exotic matter.

However, there are some quantum effects which one can use to lower the energy of empty space in a small region. The so-called Casimir effect is one such method, and it might in principle be used to construct a warp bubble. First estimations suggest that, in order to produce a warp bubble that could carry a small ship, one would need more energy than we have in the entire visible universe, though. But that might actually be a construction problem – different arrangements with warp rings and warp

tori, which need much less energy, have been proposed.

Warp Lanes through the Universe

But that is not the greatest problem: The negative energy would need to travel in front of the space ship, at a speed faster than light, so that it could bend space and time so that the vessel could then travel in its wake. But the exotic matter itself would not have a warp bubble, it would have to travel by conventional methods. As it has been stated "One needs an Alcubierre drive to build an Alcubierre drive." So is this idea of faster-than-light travel thwarted?

Well, not quite: It has been shown that one could, in fact, arrange the exotic matter along the travel path of the space ship, before the trip begins. This still kind of defeats the purpose of the whole exercise, though. Assume you decided you wanted to travel to a far away star system. Say, you chose Rigel, which is about 900 light years away, for your destination ([6]). Then, in order to build your warp drive lane from Earth to Rigel, you'd have to start

---

[5]: *"Exotic Matter" on page 271*
[6]: *"E=mc² on page 237*

placing the exotic matter along the track by conventional means of travel. So, in the end, it would take you at least 900 years (much longer if your space-ship is not super fast) to build the warp lane. In the end, you could use it to drive another ship from Earth to Rigel in no time – but the planning needed to be done quite a long time in advance. So one could, maybe, organize a network or travel lanes throughout the universe. But there would be no spontaneous visits to a place without a warp lane previously built to.

Also, once on the track, there would be no stopping or turning around, once the trip had started. Remember that the inside of the warp bubble is completely cut off from the rest of the universe. The astronauts could never change the course of the track, or decide they wanted to stop, because for the duration of the trip, they are completely trapped inside the bubble.

There is yet another point which would probably make traveling via Alcubierre drive practically unfeasible – or at least very dangerous: The universe itself is not completely empty. Rather, it is filled with gas, dust, and radiation, even if its density is extremely low. Nevertheless, a ship with an Albucierre drive would travel through the universe, picking lots and lots of matter up along the way. Now the inside of the bubble is perfectly shielded from the outside, so when the ship would fly through a star – that would rather be a problem for the star than the ship! Still, all of the accumulated matter would be kept and carried along for the ride – until the ship arrives.

When the vessel were to reach its final destination, the warp bubble becoming weaker and eventually breaking down, the interior and the rest of the universe would come into contact again. By then, the interstellar matter would be compressed from several thousand light years, down to a region only a few yards wide. That would make it hot. It is actually not quite clear who would suffer more: The ship, arriving with enough matter to form a small star just in front of it, or whoever is in its flight path then.

With Warp Technology, the Roadkill Might Kill You!

One possibility would be that all of the matter would be released in an extremely high energetic beam, blasting away everything in its wake. If that were the case, it would probably be a good idea to not directly aim at the planet one wanted to visit – it might not be there any more, when one arrives.

Another possibility would be that the matter would simply contract under its own weight – depending on how much it had been compressed during its flight (that would depend on the details of the warp bubble's shape). It should form a black hole then, if it has not done so already during the trip, and suck the ship, and possibly its destination planet, out of the universe.

All of this indicates that warp travel, if it ever becomes possible, would be not without its risks, to put it mildly. By 2016, when this book is written, a serious proposal for a fully functioning warp drive is still completely lacking.

Supersymmetry
A Beautiful Solution to Many Problems

Our world is full of symmetries. Human beings are (almost) symmetric: our left and right side look very similar. We say that we are mirror-symmetric along a vertical axis in the center of body. Another nice symmetry is realized in snowflakes. You can reflect them at different axes or points or rotate them with certain angles and you will get exactly the same structure again.

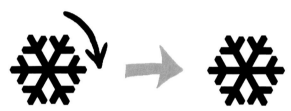

Not only are the objects that scientists describe symmetric in many ways, but so are their equations. In the chapter about conservation laws (↗[1]) we got to know several symmetries. Translation symmetry (or invariance) allows us to move in another direction, performing an experiment again and getting exactly the same result (given that the environmental conditions do not change). We also learned about other invariances like rotational invariance and time invariance. There are even symmetries that concern particle properties, such as the charge symmetry. It states that you can observe an interaction between charged particles, replace all charges with their anti-values (positive charges become negative and vice versa) and get the same reaction. Take two charged electrons repelling each other. If you invert their charges and turn them into positrons, they will

repel as well, also with the same strength as the two electrons would do.

It's the Spin That Makes You Unique

Next to all these nice symmetries there are also properties which cannot be turned into their coun-

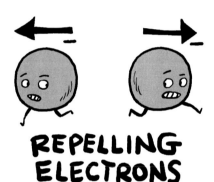

REPELLING ELECTRONS

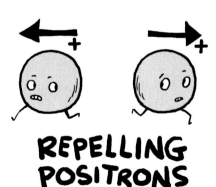

REPELLING POSITRONS

↗[1]: *"Conservation Laws" on page 51*

terparts. If you take a look at all the particles in the Standard Model of particle physics (↗2) you will notice that all the matter particles carry a spin (↗3) of 1/2. In contrast to that, all those particles that mediate interactions, the gauge bosons, carry a spin of 1. And then there is also this special particle with spin 0, the Higgs boson (↗4).

There is absolutely no way to say: "Okay, we could simply swap the spins of – let's say – an electron and a photon." One fundamental difference between these different spins is that you can put as many objects with the same properties at the same place, as long as they have an integer spin, such as our bosons. This is why there is no limitation in the number of photons in a box, for instance. This is not possible for fermions – particles with half-integer spin. This phenomenon is also known as the "Pauli exclusion principle" in atomic physics. Two electrons cannot exist at the same place if they do not vary in at least one quantum number (a quantum mechanical property).

Make the World Supersymmetric!

But what if there is a symmetry between fermions and bosons? What if for each type of fermion – so for each type of matter particle – a corresponding superpartner type would exist, having exactly the same properties but an integer spin? And the same for the bosons? Theorists have come up with such

an idea. This new symmetry has the name "Supersymmetry" – not a modest name, is it?

Obviously, nature is not supersymmetric. None of our fermions has a supersymmetric partner. Otherwise we would have seen them already. But what if the symmetry was not exact, but somehow broken? Several of nature's symmetries are broken: Many of the snowflakes are not fully symmetric, but usually we always take pictures of the nice, symmetric ones. And also human bodies are not fully symmetric. The heart is slightly on the left, the stomach as well, the liver on the right and sometimes even the outer parts are slightly asymmetric such as the nose of one of the authors (let us call him Mr. L. to keep his identity anonymous). So we could also have a broken supersymmetry. This would lead to superpartners of our particles which have masses that are much, much higher. So an unbroken symmetry would lead to the partners having the same masses (which we have not observed), while a broken one results in different masses while still every particle type would have its superpartner.

Just an Idea, but a Good One

So far we have only talked about an idea coming from theorists. But what would it be good for? Well, supersymmetry would simply be beautiful! Whatever

Supersymmetric

Electron   **Selectron**

↗2: "Standard Model of Elementary Particles" on page 213

↗3: "Spin" on page 187

↗4: "The Higgs Mechanism" on page 241

makes the formulas describing our world more symmetric is very satisfying. Just think of the Maxwell equations which tell us that electric and magnetic phenomena are basically the same. But supersymmetry is about more than just beauty. It could help us solve quite a few problems for which our Standard Model has no good answer. The first one concerns the unification of forces. Maxwell (not our dog character, but the human physicist James C. Maxwell) unified the electric and the magnetic force to the electromagnetic force. The physicists Glashow, Weinberg and Salam managed to show that the electromagnetic force and the weak force (↗5) can be unified to the electroweak force. But so far, a unification of the electroweak force and the strong force (↗6) could not be established. Supersymmetry could help to establish a theory in which all fundamental forces (without gravity) are unified. But not only that: Some of the supersymmetric particles that are predicted are good candidates for the dark matter (↗7), which is observed in the universe but can not yet be explained.

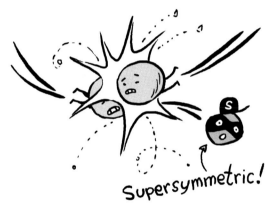

Protons collide, producing a neutralino:

Supersymmetric!

The Hunt for Supersymmetry

As supersymmetry would help to solve all of the quoted problems, experimental physicists try their best to find hints that confirm the existence of su-

persymmetry. Currently, supersymmetry is not in a good position in that respect. So far, no supersymmetric partner particles have been observed yet. Physicists try to produce supersymmetric particles in high energetic collisions of particle accelerators (↗8). A popular candidate is the "lightest supersymmetric particle" (LSP). It could be stable, neutral, and barely interacting. That's why it would not leave a signature in a particle detector (↗9). But in case the LSP is produced together with other particles one can measure all these other particles and then check that something is missing. And by checking that this missing part cannot be explained by any particle of the Standard Model (such as neutrinos, ↗10), LSP particles could be found. But so far, searches have been unsuccessful. And there are several possible explanations for this. First the most obvious: because the theory of supersymmetry is wrong. No supersymmetry, no supersymmetric particle to be found. Another explanation is that the supersymmetric partners are simply too heavy to be discovered. The heavier a particle is, the more energy is needed in particle accelerators to produce it. This is why particle accelerators as the LHC (↗8) keep increasing their energies. The search will continue, and maybe even bigger accelerators will be built. Simply because supersymmetry is too good to be just an idea.

↗5: "The Weak Interaction" on page 233
↗6: "The Strong Interaction" on page 229
↗7: "Dark Matter" on page 133
↗8: "Particle Accelerators" on page 249
↗9: "Particle Detectors" on page 253
↗10: "The Neutrino" on page 209

"It needs further study, but we've confirmed the existence of the strings we'd have to pull to get funding for it."

String Theory
A Way to a Theory of Everything™

In the sixties and seventies of the twentieth century, experimental particle physicists discovered more and more particles, using ever larger detectors (↗¹). It was a golden age for physicists, but the sheer complexity of the vast family of particles baffled them. Was there any order in the chaos?

Order in Chaos: Unifying the Particles

As it turned out, there was: The physicists eventually realized that there was only a small number of elementary particles, neatly organized in families, forming what is nowadays called the Standard Model of particle physics (↗²). All other particles they had seen in their detectors were composites, constructed out of the Standard Model particles. That way, one was even able to predict the existence of further particles, which one had not seen in the detectors before, but which had to exist. Their discovery (such as for instance the Ω^{2+} (read: "Omega-two-plus"), consisting of three up-quarks) was a great success for particle physics.

Even More Order: Unifying Three of the Four Forces!

But there was even more order than the Standard Model would let on at first: It seemed like four fundamental forces governed the interaction of all matter – the weak force, the strong force, the electromagnetic force and gravity. However, it experiments performed at higher and higher energy, two of them – the electromagnetic and the weak force – seemed to become more and more similar. And indeed, calculations showed that there is a point at which the two become part of the same force – this is called "electroweak unification". The last time the energy density in the universe was high enough was very shortly after the Big Bang (↗³). Nowadays the two seem different and separate, but they are really just two sides of the same coin.

Something similar seems to be the case with the strong force: At a very high energy (far beyond the reach of any of our current accelerators) it seems to become very similar to the electroweak force. The two combine in what is called "Grand Unification" – very shortly after the Big Bang that seemed to have been the state of the art in the universe (a precise mathematical formulation of a GUT is still work in progress, though, ↗⁴).

Since this has been known, it was the dream of particle physicists that they would one day find a way to describe all matter and all four forces – even gravity – with one unifying framework. This is what is called the "Theory Of Everything", and the thought of it is indeed very appealing. Currently, we are very

---

↗¹: *"Particle Detectors" on page 253*
↗²: *"Standard Model of Elementary Particles" on page 213*
↗³: *"Timeline of Our Universe" on page 100*
↗⁴: *"Supersymmetry" on page 299*

far from understanding how that would work, given that we do not even have a way to describe the gravitational interaction between elementary particles (↗[5]).

A Theory of Everything?

But, by the end of the seventies, and beginning of the eighties of last century, there seemed to be a silver lining: while playing around with formulas describing the gluon interaction along a line between

two quarks, theoretical physicists found out something intriguing. If one thought of particles not as (quantum) points, but as extended objects – more precisely, lines – then there would be a host of interesting consequences!

First of all, these quantum lines could either be closed, forming little loops, or open, flying around, wiggling their little open ends. Furthermore, they could – and would – oscillate, in all kinds of different frequencies. It was at this point that they were called "strings", for obvious reasons.

All Elementary Particles as Strings?

Now, that in itself is definitely interesting, but what does it have to do with the unification of the four fundamental forces? The answer is simple: different string oscillations correspond to different types of elementary particles!

A string can vibrate in different ways, and it turns out that each vibration mode (each playable note on the string, to keep the musical analogy) lets the string behave quite differently. Depending on the exact frequency, it can either be like an electron, or like a photon, or like one of the other particles that appear in the standard model of particles. Even better: there is a certain oscillation which makes the string into a massless particle with spin 2 (↗[6]) – exactly as one would expect the graviton, the tentative

↗[5]: *"Quantum Gravity" on page 279*
↗[6]: *"Spin" on page 187*

exchange boson of the gravitational interaction, to be. So, not only does string theory suggest that all particles in the universe are made up of differently vibrating little strings, it could also be the case that string theory could describe all of the four forces in a unified way. The star candidate for the Theory Of Everything was born!

String Theory Feynman Diagrams: Much Nicer Anyway!

There was another benefit of treating the elementary particles as (one-dimensional) strings instead of (zero-dimensional) points: The Feynman diagrams are much better behaved (↗7)! The Feynman diagrams of the standard model describe different interactions between elementary particles. Particles are depicted as lines following a straight line – until they undergo an interaction. Then they split off into several other particles, which whizz off in different directions. To compute the probability of such a process taking place, one needs to evaluate some complicated formulas, which give a chance of "infinity %" as answer if one does not treat them carefully.

A Feynman diagram for string theory looks quite differently: Since in Feynman diagrams the time pa-

rameter goes from the left to the right, a closed string moving through time looks like a wobbling tube. An interaction of a string (for instance, a decay of one into two strings) would be like an amoeba splitting into two. The corresponding Feynman diagram would then show one tube coming in from the right, splitting off into two tubes, leaving the diagram to the left.

It is the fact that there is no hard "pinching off" in the string Feynman diagram, which makes the corresponding formulas much better behaved than in the case of point-like particles. There, the point where one particle splits off into many is one where the formulas become quite complicated – the string diagram, however, is given by one, smooth, surface. The formulas are also much better behaved: they do not result in "infinity %" that often!

So string theory is indeed a very promising, and tantalizing, subject. Could it be that it provides a Theory Of Everything, describing all matter and forces in a unified way? Well, unfortunately not all is roses and sunshine with string theory, it also has some issues, which spoil the fun. Some of these we will treat in another article (↗8).

point-like particles

1-Dimensional strings

time

↗7: *"Feynman Diagrams" on page 225*
↗8: *"Extra Dimensions" on page 307*

Extra Dimensions
Tiny Spaces Hiding Out

In the eighties and nineties of the last century, the thought that particles could be little vibrating strings instead of point-like particles, gained more and more traction (↗[1]). Different types of particles were nothing but different oscillation modes of the small, one-dimensional lines consisting of pure energy. Also, it seemed like the theory was quite "rigid", in the sense that one could not change details of the theory easily. This was celebrated by physicists, as the great predictive power of the theory: there were not many different ways to do string theory, but only very few, with very specifically fixed parameters. And these few (five, to be precise) different string theories seemed to be related by symmetries, i.e. appeared to be one and the same theory, just in five different formulations.

String theory seemed to have the potential to unify all known particles and interactions, to build the foundation of a Theory Of Everything.

Wanted: Six Extra Dimensions

There was one caveat, though, which was apparent immediately. It only worked, when space-time was ten-dimensional (one time- and nine space-directions)! This is very much at odds with what we experience in our daily life: isn't it clear that there are one dimension of time and three dimensions of space, rather than nine? But there was no way around it: with any other number of dimensions, string theory would just become inconsistent. This was one of its rigid features: one could not just write down an equally-well working string theory in four space-time dimensions. It was either ten, or nothing!

Now what to do in this situation? Was string theory just plain wrong? But in many other respects, it seemed too good to be true, so physicists were not ready to give it up just yet. And indeed, there was a way out of the dilemma!

How to Hide Extra Dimensions: the Garden Hose Analogy

You see, the ten dimensions only need to exist, they do not have to be large. Instead, they could be curled up so much that one could not see them with the naked eye.

The idea of "curled up dimensions" can be easily visualized by a garden hose. If you look at your garden hose from very close up, you can see that it is a long tube. Imagine a small insect, such as a

↗[1]: "String Theory" on page 303

mite, crawling around on it. The mite has two independent directions it can scuttle: along the length of the hose, and around its circumference. Because a mite is quite small compared to the thickness of the hose, it will probably not even realize that one of the two directions is curved in on itself. If it leaves a marker somehow, it could realize that it would return to where it started after a short time of crawling sideways around the hose. Otherwise it would just see the two dimensions, and not think much about it. So, from close up, the (surface of the) hose is two-dimensional.

Form far away, however, it is not. Imagine you flying in a helicopter over the suburbs, and peering down, seeing someone watering his garden with a hose. If you have good eye sight, you might see the hose as a very thin line, with water coming out on one end. It would certainly look completely one-dimensional to you. There would be no way for you to see a mite crawling sideways on the surface of the hose, it would simply be too far away! So, from far away, a garden hose is one-dimensional.

Another way to say all this is that the surface of the hose is two-dimensional, but one of the dimensions

is large, and the other one is curled up. If you don't have the ability to look at it closely, you will only see the large dimensions, not the small, curled up ones. The situation with string theory is imagined to be very similar: out of the nine dimensions of space, six are tightly curled up, and only three are still large. And the difference is so enormous that it seems to us as if the world was four-dimensional, instead of ten-dimensional.

"You Must Be at Least This Energetic to Ride"

String theorists assume that the six additional dimensions are rolled up really small – they call these the "compactified directions". So there are three directions in which the universe is large, and six directions in which the universe is really, really small.

But even if these additional compactified directions are small, shouldn't we be able to see them? We should still be able to walk into these directions, it's just that we'd immediately return to the point had started, right? Well, not quite. Because of quantum physics, you would need an enormous amount of energy to go into one of the six compactified directions. The reason is that the wave function of a particle is, as the name suggests, a wave ($\nearrow$2). If a wave travels along a direction in which it has as much room as it wants, then it can have any wavelength. But if it wants to travel into a compactified direction, then it would be able to travel around the short direction and come back, interfering with itself. In order for this interference not to be destructive, it would need to have a very short wavelength – at most as long as the radius of the universe in that direction.

So a particle traveling along one of the short, compactified directions would need a really, really short wavelength – about as short as the size of that dimension. And a short wavelength means a large energy ($\nearrow$3). Given that many string theorists assume the size of these extra dimension to be around the Planck length of about 10^{-35} m, the energy required for a particle to move in any other than the three dimensions that we all know and love, would be so large that it could never be created in one of our particle accelerators ($\nearrow$4). Shortly after the Big Bang particles would have had enough energy for that, but nowadays it would be practically impossible to find a particle in the universe with that much energy.

Some physicists working in string theory are entertaining the idea that these extra dimensions are, in fact, large (and by large, they mean "large in comparison to the Planck length, but still a lot smaller than an atom"). If that were true, one might be able to see them in particle accelerators like the LHC ($\nearrow$4).

How to Curl up Precisely? Too Many Options!

These extra dimensions in string theory have led to one of its greatest crises, though. You see, if you have only one extra dimension which you want to make small, there is essentially only one way of doing it. You can only curl it up in the way the garden hose has one curled up dimension. But if you have six of them, there are many possibilities. Nobody knows precisely how many – people have tried to estimate the different ways this can be done, and some have come up with the number of 10^{500} – that is a 1 with five hundred zeros! And each of these different possibilities leads to very different theories, with different types of particles, different constants of nature. And it seems absolutely impossible to guess which is the right one, or even if there is one which describes our universe correctly.

So the whole predictive power of string theory seems to be lost: at first, it appeared as if the rigidity of string theory allowed only for one way the universe could be. Now it seems that there are billions of billions of different string theories – all with differently compactified extra dimensions – and no good way of figuring out which is the right one. Maybe this problem will be solved in the future, but at the point of writing this book (2016), it appears as if one is not much closer to a Theory Of Everything than one was thirty years ago.

$\nearrow$2: *"Wave-Particle Duality" on page 143*
$\nearrow$3: *"Light" on page 7*
$\nearrow$4: *"Particle Accelerators" on page 249*

Many Worlds
The Cat Is Alive in Another Universe

The rules of quantum physics are fundamentally probabilistic in nature. Every process that is possible only happens with a certain probability. This probability can be computed precisely by quantum mechanics – but which of the possibilities actually happens, cannot be predicted. But, in fact, it is even stranger: As long as nobody looks, all possible outcomes of a process are still realized!

The Superposition: When Does It Stop to Exist?

Imagine a radioactive atom, for instance a uranium 238 atom. It is radioactive, with a half-life of about 4.5 trillion years. So if you wait for that time, what will happen? By the rules of quantum mechanics, the state of the system after 4.5 trillion years will be a superposition of a uranium atom on the one hand, and a thorium atom and an alpha particle on the other hand (↗1). The probability for either possibility will be 50%.

By the so-called "Copenhagen interpretation" of quantum mechanics, either possibility will only be realized when you make a measurement, i.e. look whether the uranium atom has decayed or not. Only at that instant will the state of the system change – if you measure a uranium atom, the state will be a uranium atom with 100% probability. If you measure that the atom has indeed undergone a decay, then the state will be a thorium atom and an alpha particle, with 100% probability. This measurement process will change the state, at the time of the measurement.

The Copenhagen Interpretation: Collapse of the Wave Function

This interpretation of the measurement process is certainly slightly problematic. For once, it claims that the wave function of the state changes everywhere, instantly. In a world where we know that any physical process can only happen at most as fast as the speed of light (↗2), this notion seems a bit strange.

But even worse, it makes a clear distinction between the quantum system on the one side, and the measurement apparatus (or rather, the scientist making the measurement) on the other side. The latter is treated completely classically, that is to say, no quantum effects of the measurement apparatus

↗1: *"Alpha, Beta and Gamma Rays" on page 171*
↗2: *"Relative Space and Time" on page 117*

are taken into account! Most notably, a measurement device can never show "the atom has both decayed and not decayed" as a result.

But both the measurement device, as well as the scientist, also consist of atoms, which should follow the rules of quantum mechanics, right? So, what makes one quantum and the other one not quantum? Where is the boundary between the quantum and the classical world? Schrödinger found this notion so strange that he thought up ingenious ways of torturing cats to prove his point: this could not be the whole story! (This was just a thought experiment, he didn't do it to real cats! ↗³)

This seems even more strange, since it is possible to build "quantum erasers": It is possible to make a measurement, and then carefully make sure that the result of the measurement never reaches the scientist, but rather is destroyed. In that case, the state is still in a superposition – so it seems that it is not the measurement process which destroys the superposition, but rather the fact that anybody knows about which of the two possibilities is realized!

For many reasons, some of which we have just described, the Copenhagen view of quantum mechanics does not seem appropriate anymore – even though it is a good description of what happens, if one neglects the theory of relativity, and does not

think too hard about the philosophical consequences.

Everything Is Quantum: Entangling Atom and Measurement Device

Well, what would happen if there simply was no boundary between the quantum and the classical world? Let's think that through: the quantum state is in a superposition of "uranium atom" and "thorium atom and alpha particle". Now think of the measurement apparatus as one giant quantum mechanical system, with two possible quantum states: "device reads: uranium has decayed" and "device reads: uranium has not decayed". What happens if the measurement device is used to see whether the atom has decayed or not? Will it also be in a superposition of its two possible states?

No, it turns out it won't. Rather, after the measurement process the combined system of "uranium atom + measurement device" will be entangled! (↗⁴) The combined system will be in a superposition of the following two possibilities: "uranium atom + device reads: uranium has not decayed", and "thorium and alpha particle + device reads: uranium atom has decayed". The two possible outcomes of the measurement will still be represented in a quantum state, which includes both atom and measurement device. Needless to say, as soon as the scientist reads off the measurement result, the system "atom + device + scientist" will be

---

↗³: *"Schrödinger's Cat" on page 155*
↗⁴: *"Entanglement" on page 191*

in a superposition of "uranium atom + device reads: uranium has not decayed + scientist has a uranium atom" on the one hand, and "thorium atom and alpha particle + device reads: uranium has decayed + scientist wonders where he'll get a new uranium atom" on the other hand.

The Universe Splits Off

So in a sense, both outcomes of the measurement are still happening, at the same time! But something else is happening as well: Because the measurement device (and the scientist) consist of so many, many different quantum particles themselves, which interact a lot with each other, the wave function will not collapse – but it will decohere. This decoherence means that the two possi-

bilities will, after a quite short time, evolve very much independently, and not interact with each other anymore. It is as if the universe had split into two parts: one where the uranium atom has decayed, and one where it hasn't.

This interpretation has been originally brought up by Hugh Everett in 1957, and later been popularized as "many-worlds-theory", or "many-worlds interpretation of quantum mechanics". In short, it says that, whenever a measurement of a quantum is performed, the universe splits off into as many parts as there are parts in the superposition. In each

universe, one result of the measurement is realized. And after that, each part of the universe (each "world") evolves separately from the others, without possibility of communication between them.

Parallel Universes for Everyone!

Needless to say, the many-world interpretation of quantum mechanics has been subject to heated discussions among physicists throughout the decades. In particular the fact that the exact nature of decoherence is poorly understood, is reason for repeated criticism. Also, it seems that this interpretation would be time-asym-

metric: the universe can split into different branches, but there is no merging of universes, while the rules of quantum mechanical evolution are time-symmetric (while the measurement process in the Copenhagen interpretation is not time-symmetric either, of course).

This will probably be a point of discussion for some time. And, of course, it has fueled, and will fuel many fantasy- and science-fiction stories! Which author can say no to scientifically sanctioned parallel universes?

The End of the Universe
And Then?

We know quite a bit about how the universe began: about 13.8 billion years ago, the universe exploded into existence at the big bang (↗1). It was very small and hot at that time, but has expanded and cooled off ever since. But how will the universe end? Will it expand forever? Or collapse back to a point? Or is there another possibility entirely?

To start, we should admit: nobody knows how the universe will end (despite there being a surprising number of apocalyptic prophets who predict the end of times on a regular basis). But, according to the physical equations that we have, which describe space, time and matter to the best of our knowledge, as well as the state and shape our universe is in today, there are a few scenarios which are more or less likely to happen.

The Big Freeze

At the current time, the universe expands rapidly (↗1). And, as far as we can tell, that expansion rate even increases over time, so the universe is getting larger faster and faster!

One thing that does not increase, however, is the total amount of matter and/or energy – there seems to be just a finite amount of it. (As a side remark: In the mid-20th century, an idea called "steady state cosmology" was quite popular, which basically suggested that not only space was expanding, but also new matter was created, continuously. That model is obsolete nowadays, since it does not match the experimental data, unlike the big bang model.) This means that the same amount of energy is spread thinner and thinner, until there is not enough energy density anymore to let anything interesting happen. There won't be enough matter density to fuel new star formation (↗2). So all stars will, one after the other, go out, with only some black holes remaining (↗3). And even those will, after a long time, radiate away because of Hawking radiation (↗4). The temperature of the universe will approach absolute zero – it will freeze to death.

The Heat Death

A variant of the Big Freeze scenario is the Heat Death of the universe. This scenario rests on the second law of thermodynamics: a physical system always tends to the state of maximum entropy – in the terms of our universe this means that, eventually, all matter and radiation will be more or less evenly distributed, everywhere. Without any variations in temperature, nothing can be happening, and no life will be possible.

↗1: *"The Big Bang" on page 97*
↗2: *"Birth of the Solar System" on page 59*
↗3: *"Black Holes" on page 91*

↗4: *"Black Hole Evaporation" on page 283*

Whether the Big Freeze or the Heat Death will occur, actually depends on how fast the universe expands in the future, and whether black holes can decay or not.

The Big Rip

The theory of general relativity (↗5) allows for several singularities to occur – the black hole and the big bang being only the two most famous ones. There is another type of singularity, in which the expansion rate of the universe becomes infinite after a finite time. In this scenario, the universe suddenly expands so rapidly in such a short time that it "rips apart": Any two points in space will be accelerated away from each other infinitely fast! Everything will be ripped apart in an instant, and nothing of the universe is left.

Luckily, this can only happen if the universe is filled with some kind of mysterious force called "phantom energy", which is a special form of dark energy. If there are actually only baryonic matter, dark matter, and dark energy (as we understand it), as our current models suggest, the Big Rip singularity will not occur.

The Big Crunch

At the moment, the universe is expanding rapidly, and it appears that it will continue to do so. However, if it turns our that it is not infinite, but finite in size (one also calls this "closed" universe), and that certain predictions in quantum gravity theory are correct (↗6), then the expansion of the universe could stop at some point in the far future, and it would begin to shrink again afterwards. The end of the universe would occur at the point in which all of it collapses to a point – the Big Bang in reverse, called the "Big Crunch".

The Big Bounce

If certain quantum gravity theories are to believed (↗6), then the Big Crunch does not need to be the end – instead of collapsing to a point, the universe could collapse to a very small, but finite size. At that point, when the density becomes about as large as the Planck density (roughly the mass of our galaxy compressed to the size of a quark), gravity becomes repulsive, instead of attractive, and the universe "bounces back". It starts to expand again rapidly! A new universe is born, with a new Big Bang, starting everything from anew.

Perhaps one of the most hopeful of the scenarios, don't you think? Whatever will happen, though, it will happen so far in the future, we will probably be gone much before. Though again, who knows what the future brings?

↗5: *"The Theory of General Relativity" on page 121*
↗6: *"Quantum Gravity" on page 279*

... and then
what ?